GROLIER
FUNDAMENTALS OF SCIENCE SERIES

BASIC SCIENCE ACTIVITIES PREPARATION GUIDE

HAINES JR. HIGH SCHOOL
RESOURCE CENTER
ST. CHARLES, IL 60174

CHAPTER 2-1991

Grolier Educational Corporation
Danbury, Connecticut

© 1990 by

ALPHA PUBLISHING COMPANY, INC
Annapolis, Maryland

All rights reserved. No part of this book may be reproduced or transmitted in any form by any means electronic, mechanical, or otherwise, whether now or hereafter devised including photocopying, recording, or by any information storage and retrieval system without the express written prior permission from the publisher, with the following exception:

Permission is given for individual librarians and science teachers to reproduce the student activity pages and illustrations for home and classroom use. Reproduction of these materials for an entire school system is strictly forbidden.

Produced in the United States of America by Alpha Publishing Co., Inc., 1910 Hidden Point Road, Annapolis, MD 21401

Printed in the United States of America

First Edition

SERIES ISBN: 0–7172–7167–6

Library Of Congress Catalog Card Number : 90–82821

GROLIER
FUNDAMENTALS OF SCIENCE SERIES

BASIC SCIENCE ACTIVITIES PREPARATION GUIDE

FUNDAMENTALS OF SCIENCE

The *Fundamentals of Science* is a series of science activity books designed to address the significant changes and reforms in K–6 science education and child learning theories.

These books are written with the aim of promoting science education in elementary schools. The activities included in *Basic Chemistry, Basic Biology, Basic Physics*, and *Basic General Science*, are intended to nurture the skills of students in laboratory techniques and critical thinking. These activities also provide opportunities to master a body of knowledge in a variety of topics.

The safe, interesting, and exciting activities included in this series are intended to enhance the elementary students' interests and attitudes toward science and help them become scientists.

Fundamentals of Science also serves as a unique resource series for elementary school teachers. These books lessen the difficult task of collecting and assimilating information on many science topics. (It is generally agreed that because of these difficulties, science is often ignored in elementary schools.) The brief review of biology, chemistry and physics and general science concepts are intended to strengthen the basic scientific background of elementary teachers.

The wide variety of choices of activities in this series offer equal opportunities for the average as well as the most gifted and talented students. The selected experiments include both easy and challenging experiments.

The Teacher's Section for each student laboratory activity and teacher demonstration offers considerable background information for the teacher in understanding concepts, collecting materials, and discussion both before and after the laboratory activity. Many of the activities can be assigned as collaborative assignments with parents at home. Most of the laboratory activities utilize common house-hold materials, chemicals and simple equipment.

INTRODUCTION

FUNDAMENTALS OF SCIENCE ACTIVITY SERIES

Basic Biology Science Activities
Basic Chemistry Science Activities
Basic Physics Science Activities
Basic General Science Activities

The *Grolier Fundamentals of Science* is a series of science activity books designed to address the significant changes and reforms in K–6 science education and child learning theories

The series is written with the aim of promoting science education at the elementary schools. The activities included in *Basic Biology, Basic Chemistry, Basic Physics* and *Basic General Science* are intended to nurture the skills of students in laboratory techniques and critical thinking. These activities also provide opportunities to master a body of knowledge in a variety of topics. The safe, interesting and exciting activities included in this series are intended to enhance the elementary student's interests and attitudes toward science and help them become scientists.

Fundamentals of Science also serves as a unique resource series for the elementary school teacher. The series lessens the difficult task of collecting and assimilating information on many science topics (It is generally agreed that because of these difficulties, science is often ignored in elementary schools.) The brief review of biology, chemistry, physics and General Science concepts are intended to strengthen the basic scientific background of the elementary educator.

The wide variety of choices of activities in this series offers equal opportunities for the average as well as the most gifted and talented students. The selected experiments include both easy and challenging experiments.

The *Preparation Guide* for each student laboratory activity and teacher demonstration offers considerable background information for the teacher in understanding concepts, collecting materials and discussion both before and after the laboratory activity. Many of the activities can be assigned as collaborative assignments with parents at home and most of the laboratory activities utilize common house-hold materials, chemicals and simple equipment.

ABOUT THE AUTHORS

Nancy Coggins Lynch, author of *Basic Biology*, holds a BA in Biology, and a BS in Secondary Education from State University College at Oswego. She holds an MA from SUNY at Stony Brook. During her 15 years at East Islip High School she has taught all levels of Biology, including Advanced Placement, three different levels of Chemistry, as well as Earth Science, Marine Science, Psychology, and General Science. She has advised over 100 honors students in their Biology and Chemistry research projects. In 1984 she was selected to be the Biology teacher in the Academy, a school within a school program initially funded by a Carnegie Grant.

For four years Ms. Lynch served on and also chaired the organizing committee for Science Explorations for Suffolk County, and has served as secretary for the Suffolk County Science Teacher's Association.

In addition, Ms. Lynch is an adjunct professor at SUNY, Stony Brook, and a consultant to the Office of Educational Programs at Brookhaven National Laboratory where she continues to serve as a program advisor for the National Synchrotron Light Source Honors Research Program. She is also a demonstrator for the Saturday Junior High School Science Program at BNL.

Professor C. K. Krishnan, author of *Basic Chemistry* and *Basic Physics*, had his early education in India, where he received a gold medal for his Ph.D Thesis in Physical Chemistry. He worked in the Bhabha Atomic Research Center in India for 10 years. He moved to United States in 1967 and had been teaching at East Islip High School for the last 20 years. He has also worked at State University of New York at Stony Brook and Brookhaven National Laboratory during the last 22 years. He is now a visiting Professor of Chemistry at State University of New York at Stony Brook.

Professor Krishnan has done extensive research in solution thermodynamics, statistical mechanics and photochemistry. He has contributed to more than 45 research papers and chapters in 4 books on solute-solvent interactions. He is also one of the two writers of the revised New York State Regents Chemistry Syllabus.

Professor Krishnan has taught Advanced Placement Chemistry, New York State Regents Chemistry and Physics, Non Regents Chemistry and General Science at East Islip High School. He has also taught freshmen General Chemistry and Elementary Chemistry, and Graduate Courses for teachers, Methods of Teaching Chemistry in Junior High Schools, Methods of Teaching Chemistry in High Schools, Demonstrations in Chemistry, and Chemistry for Elementary Schools at State University of New York at Stony Brook.

Professor Krishnan has recently established a Chemical Education Resource Center at Stony Brook and is actively involved in improving chemical education on Long Island. He also gives "Magic of Chemistry" demonstration shows under the auspices of Stony Brook. He also gives 3 different week-long summer workshops, "Parents-Children Partnership In Chemistry" for students, K through 6, and parents.

Professor Krishnan has received numerous awards including the 1984 Presidential Award for Excellence in Teaching Science and Mathematics, from the President of the United States, Chemical Manufacturers Association's National Catalyst Award (1989) and American Chemical Society's Nichols Award (1984).

Eugene Kutscher, author of *General Science*, is Science Chairman and Coordinator of Science Research for the Roslyn Schools, Roslyn, New York. Their science department, nationally recognized for excellence, emphasizes scientific processes through an independent research program as well as in its classroom activities. The philosophy of educational excellence in science is an actuality at Roslyn, and the activities incorporated into this books reflect that standard. They are exciting, easy to comprehend, and yet lend themselves to concept development rather than to rote learning.

Mr. Kutscher holds a B.A. and M.A. in Physics from the City University of New York. He has spoken on the topic of "How To Start A Successful High School Science Research Program" at the New York State School Boards Association, the annual meeting of the Science Teachers Association of New York State, and under the sponsorship of the National Science Supervisors Association, at the annual conference of the National Science Teachers Association.

In addition, Mr. Kutscher lectures on nuclear and alternative energies, is active in Science, Technology and Society groups, is on the National Board of Directors of Zero Population Growth, and represents the National Science Supervisors Association at the Alliance for Environmental Education.

This year, he again conducted a National Science Supervisor Association endorsed workshop on high school science research and participated in a panel on the same subject at the National Science Teachers Association annual meeting. He has presided and spoken about population, energy and other science-technology-society issues at national and regional NSTA meetings, professional growth day workshops, on television's "Straight Talk" and at the Edison Electric Institute's meeting for utility educators.

Mr. Kutscher has been a science educator for twenty-two years, teaching physics, astronomy, biology, chemistry, earth science and mathematics in all grades from seven through college.

Table of Contents

Basic Biology Reseach Activities

1. Is It Alive?
   Using Methylene Blue To Tell Whether An Object Is Alive Or Not 3

2. Can You Make An Egg Get Heavier Or Lighter
   Without Breaking The Egg?
   Osmosis 5

3. What Is Inside An Egg Shell?
   Parts Of A Raw Egg 8

4. How Much Vitamin C Is There In Your Food?
   Testing For Vitamin C 9

5. Where's The Fat?
   Testing For Fat In Food 11

6. Where's The Starch?
   A Test For Starch Using Iodine 13

7. What Makes Orange Kool Aid Orange?
   Chromatography To Detect Artificial Coloring In Foods 16

8. Cabbage Magic: How To Change Red Cabbage Juice
   To A Pink Or Green Color
   Red Cabbage Indicator For Acids And Bases 19

9. How Acidic Is It?
   Using Red Cabbage Indicator To Determine
   Which Is The Most Acidic Substance 21

10. Do You Eat Food That Contains Bacteria?
    Detection Of Bacteria Using Methylene Blue 24

11. What Do You Get When You Make Thin Slices Of Carrots And Celery?
    How To Make A Cross Section 26

12. How Does Water Move Up A Plant Stem?
    Transport In A Plant Stem 28

13. The End Of The Line?
    Transpiration In Plants 30

14. Seed Packages
    Parts Of A Bean Seed 32

15. What Part Of A Seed Stores Food For A Baby Plant?
    Starch In A Bean Seed 33

16. Seed Races
    Which Seeds Germinate And Grow The Fastest? 35

17. At What Temperature Do Seeds Sprout?
    Seed Germination Vs. Temperature 37

18. Can You Grow A Plant Upside Down?
    Geotropism 39

19. Can The Roots Find The Water?
    Hydrotropism 41

20. Why Do Plants Grow Towards Light?
    Phototropism 43

21. Why Does A Plant Need Sunlight?
    Starch Production And Storage In A Leaf 45

22. Can You Grow A Plant Without Soil?
    Hydroponics 48

23. Can You Grow A New Plant Without Using A Seed?
    Vegetative Propagation 50

24. She Loves Me, She Loves Me Not.
    Can You Name The Parts Of A Flower?
    Parts Of A Flower 52

25. Can You Catch A Falling Ruler?
    Measuring Reaction Times 54

26. Listen To Your Heartbeat
    Making A Stethoscope To Measure The Rate Of Your Heart 55

27. **Race Against Your Heart**
    Can You Work As Well As Your Heart Can? 57

28. **Disappearing Act**
    Finding Your Blind Spot 59

29. **Catch A Glimpse**
    Measuring Peripheral Vision 61

30. **The Eyes Have It**
    Depth Perception 62

31. **Pupils That Don't Go To School**
    Reflex Reactions Of The Pupils Of Your Eyes To Light 64

32. **What's The Best Way To Eat A Lollipop?**
    Find The Sweet Taste Receptors On Your Tongue 65

33. **How Sensitive Are You?**
    Measuring The Accuracy Of The Sense Receptors
    In Different Parts Of Your Body 67

34. **Are You Full Of Hot Air?**
    How To Measure Your Lung Capacity 69

35. **Inhale, Exhale, Inhale, Exhale**
    Using Limewater To Test For Carbon Dioxide In Your Breath 71

36. **Can You Do It?**
    Experiments In Balance 73

37. **Can You Bend A Bone Or Bounce An Egg?**
    Using Vinegar To Remove Calcium Compounds 75

38. **Can You Make Your Own Yogurt?**
    Making Yogurt From Milk 77

39. **There's A Fungus Among Us**
    Growing Mold 79

40. **Yeast Power**
    Using Yeast To Make Bread Rise 81

41. **Making Friends With An Earthworm**
    Measuring The Pulse Of An Earthworm 83

42. **Do Bacteria Live In The Soil?**
    Using Methylene Blue To Detect Soil Bacteria 85

43. Making Tracks
    Using Plaster Of Paris To Preserve Animal Tracks 87

44. What Can We Do With All Of Our Old Paper?
    Recycling 89

45. Getting To Know You
    Learning The Parts Of A Microscope 91

46. Up Close And Personal
    Using A Microscope To Observe The Letter "v" 93

47. How Cheeky Are You?
    Using A Microscope To Observe Stained Cheek Cells 95

48. Can You Make A Brick Wall Using Onion Cells?
    Using A Microscope To Observe Stained Cells
    From The Membrane Of An Onion 98

49. Green Polka Dots
    Using A Microscope To Observe The Chloroplasts In Green Plant Cells ___ 101

50. Who Lives Here?
    Using A Microscope To Observe The Different Microorganisms
    That Live In Pond Water 104

Basic Chemistry Science Activities

1. Using A Balance 109

2. Air Is Matter 110

3. Helium Balloon 111

4. Dry Ice 112

5. Volume Of A Solid And A Gas 113

6. Volume Of A Solid 114

7. Volume Of A Solid With A Regular Shape 115

8. Molecules Of Gases 116

9. Escaping Of Gas Molecules 117

10. Temperature Measurements Using An Alcohol Thermometer And A Game Thermometer 118

11. Air Pressure 120

12. Magic With A Balloon And Two Cups 122

13. Magic With Pressure Of Air In Balloons 123

14. Air Exerts Pressure 125

15. Fun With Air 127

16. Magic By Squeezing Air 129

17. Cold Air And Hot Air 130

18. Moisture In Air 132

19. Physical Changes 133

20. Melting Point Of Ice 135

21. Freezing Point Of Water 137

22. Magic With Ice Cubes 139

23. Boiling Point Of Water 140

24. Spaces Between Water Molecules In Liquid Water 142

25. Magic With "The Heat Solution" 144

26. Magic With A Cold Pack 146

27. Magic With Super–Absorbent Diapers 149

28. Absorption Of Moisture From Air 150

29. Expansion Of An Egg 151

30. Decolorization With Activated Charcoal 152

31. Milk And Food Colors 153

32. Mixing Molecules And Water In Different Containers 155

33. Mixing Food Color And Water 157

34. Mixing A Solid And A Liquid 159

35. Movement Of Molecules In Solids 161

36. Mixing Molecules In A Solid And Water At Different Temperatures 164

37. Mixing Molecules In A Solution And Water At Different Temperatures 166

38. Molecules Of Solids That Do Not Mix 169

39. Escaping Of Molecules Of A Solid With The Help Of Water 171

40. Escaping Of Molecules Of A Solid From Different Containers 172

41. Oxygen In The Air 173

42. Carbon Dioxide In Air 175

43. Chemical Changes 177

44. Water, Salt–water, And Sugar–water 179

45. Magic Of Changing Colors On Talking 181

46. Endothermic Processes 184

47. Exothermic Processes 186

48. Maximum Amount Of Gas From The Shell Of An Egg 188

49. Carbon Dioxide Gas From The Shell Of An Egg 190

50. Bleaching Of Food Colors 193

General Science Basic Activities

1. A Million Stars 197

2. Can You Describe This? 199

3. The Banana Ruler 202

4. Why Do We Need Thermometers? 204

5. How Much Space Do You Take Up? 206

6.	Sinking The Soda	208
7.	The Jello Cell	210
8.	Enzymes And You	212
9.	Your Roots: Why You Are You	214
10.	Food Chains And Food Webs	216
11.	Camouflage: The Find The Dinner Game	218
12.	The Balance Of Nature	220
13.	The Growth Of Plant Roots And Stems	222
14.	You Are The Scientist: The Grow The Best Plant Contest	223
15.	Why Should A Nice Green Leaf Turn Red Or Yellow?	225
16.	Is Black Ink Really Black?	227
17.	Messages In Invisible Ink	229
18.	Models Of Atoms	231
19.	Race Car Trials	233
20.	Let's Move The Piano: The Inclined Plane	235
21.	Let's Move The Piano: The Lever	237
22.	Pulley Tug Of War	239
23.	Interesting Inertia	241
24.	Actions And Reactions	243
25.	Getting A Lift From Air Pressure	245
26.	Collapsing Cans	247
27.	The Static Electric Magic Wand	249
28.	The Bubble Telegraph	251
29.	The Lemon Battery	253
30.	Conducting Compounds	255

31.	Simple Series Circuits	257
32.	Fun With Magnets	259
33.	Magnetic Materials	261
34.	Making Magnets Stronger	263
35.	The Speed Of Heat Transfer	265
36.	Heat And Expansion	267
37.	Recycling Paper	269
38.	Solar Collectors	271
39.	Volume, Area And Solar Homes	273
40.	The Best Light Bulbs	276
41.	Investigating Insulation	278
42.	Air Pollution And You	280
43.	Water Pollution And Water Plants	282
44.	A True Scale Model Of Our Solar System	284
45.	The Earth's Path Through Space	286
46.	A Balloon Model Of The Universe	288
47.	Science Jeopardy	289
48.	Science Bingo	292
49.	Science Tag Team Race	293
50.	An Egg Drop Contest	295

Basic Physics Science Activities

1.	Measurement Of Length	299
2.	Measurement Of Volume	301

3.	Measurement Of Mass	304
4.	Measurement Of Time	306
5.	Measurement Of Heat	308
6.	Measurement Of Force	310
7.	Measurement Of Work	312
8.	Bouncing Ball	314
9.	Bouncing Of Different Balls	316
10.	The Nature Of The Floor And The Bouncing Ball	318
11.	Force Due To Friction	320
12.	Friction And The Nature Of Surfaces	322
13.	Fighting Friction	324
14.	Reducing Friction	326
15.	The Good Aspects Of Friction	328
16.	Check Your Hugging Power	330
17.	Static Electricity	331
18.	Two Kinds Of Static Electricity	333
19.	Attraction And Repulsion Between Electric Charges	335
20.	Electrical Force And Gravitational Force	338
21.	Suspending Coins Using Static Electricity	340
22.	Suspending A Charged Balloon From Different Surfaces	342
23.	Styrofoam Cup Contest	344
24.	Identifying Water And Oil Using Static Electricity	346
25.	Electrical Circuits	348
26.	Electrical Conductors And Insulators	350
27.	Electrical Resistance Of Different Metals	352

28.	Electrical Resistance And Length Of The Metal Wire	355
29.	Electrical Resistance And Thickness Of The Metal Wire	358
30.	Series Circuits	360
31.	Electrical Current In A Series Circuit	362
32.	Parallel Circuits	364
33.	Electrical Current In A Parallel Circuit	366
34.	Making A Light Bulb	368
35.	Magnetic And Nonmagnetic Objects	371
36.	Magnetic Attraction Through Other Materials	372
37.	North Pole And South Pole Of A Magnet	374
38.	Magic With Magnets	376
39.	Magnetizing Other Objects	378
40.	Making A Compass	380
41.	Strength Of A Magnet	382
42.	Magnetic Field Around A Magnet	384
43.	Magnetic Field Between Two Unlike Poles	386
44.	Magnetic Field Between Two Like Poles	388
45.	Mapping A Magnetic Field Using A Compass	390
46.	Magnetism From Electricity	392
47.	Making An Electromagnet	394
48.	Number Of Turns In The Coil Of An Electromagnet	397
49.	The Amount Of Current In The Coil Of An Electromagnet	400
50.	Use Of Permeable Materials To Strengthen An Electromagnet	403

Basic Biology Science Activities

by

Nancy Coggins Lynch

TEACHER'S GUIDE
IS IT ALIVE?

**Using Methylene Blue
To Tell Whether An Object Is Alive Or Not**

GOALS:

1. To observe that the color of methylene blue solution changes.

2. To identify the control in the activity as the methylene blue solution that has nothing added to it.

3. To observe that the color change in the methylene blue solution was caused by the living organisms.

4. To infer that living things are different from nonliving things due to the color change that the living things caused in the methylene blue solution.

SAFETY:

Any time you or your students handle chemicals, you should wear safety goggles. An apron is suggested to protect your clothing from the methylene blue solution. Stress that your students should immediately notify you in the case of spills or breakage. Never handle broken glass with your bare hands, use paper towels or pan and brush. It is best to have a separate container for disposal of broken glass to prevent cuts when other people empty the container. Always wash your hands after working with chemicals.

PREPARATION AND HINTS:

1. This activity can be done in groups of 4. Each student in the group should set up one of the 4 containers.

2. Place methylene blue solution in several small containers. Label each container with the words "Methylene Blue Solution". The solution can be dispensed with eye droppers. You can also use small plastic containers that dispense liquids, a drop at a time. (Small contact lens bottles work well. These containers should not be used for any other purpose after they are used for methylene blue. They can be capped, stored, and used in other activities that require methylene blue.)

3. Place the oil in several containers. Plastic containers that have small openings (like contact lens wetting solution containers) reduce the number of spills, are easy to use, and can be used for storage.

4. Obtain aquatic snails from the pet store, or by collecting them from a local body of water. You will need two snails for each group of students.

5. Place seeds in a container and cover them with water for 24 hours. (This will cause the seeds to begin to germinate. The germinating seeds will cause a color change in the methylene blue.) You will need 6 – 12 seeds for each group of students.

6. Collect marbles or small stones. You will need two for each group of students.

7. Set up a supply table. Neatly arrange the containers in multiples of four. If you are using test tubes, place them in test tube racks. (You can make them by cutting 4 – 6 test tube size holes in the side of a clean, empty cardboard milk container. The test tube rack can be saved for future use.) Place labeling supplies (crayons or markers and masking tape) next to the containers. Place the seeds in a large, flat dish or tray (it is easy for several people to get seeds simultaneously from this kind of container) that has a little water at the bottom. Place snails in a large, flat dish or tray that has enough aquarium water to cover the snails. Place the marbles or stones in a flat box or container. Arrange the small containers of methylene blue and oil neatly. (The neater the supply table is, the easier it will be for your students to find materials, and to return them neatly.)

8. Label the areas of the supply table. Your students will return the materials more neatly if they know exactly where to put the containers, oil, methylene blue, marbles, and snails.

9. Provide a waste container for the methylene blue and oil. The oil could harm the drains. Seal the container before you discard it.

10. Provide paper towels and soap for cleaning up the containers (if they are to be saved), work surfaces, and hands.

11. Provide a garbage can for paper disposal.

12. Discuss the procedure with your students.

TEACHER'S GUIDE
CAN YOU MAKE AN EGG GET HEAVIER OR LIGHTER WITHOUT BREAKING THE EGG?
Osmosis

GOALS:

1. To predict the circumference and mass of an egg.
2. To measure the circumference and mass of an egg.
3. To compare the predicted values for the circumference and mass with the actual values.
4. To observe that the size of the egg changes when placed in water or corn syrup.
5. To infer that when an egg is placed in water, water moves into the egg, making it larger.
6. To infer that when an egg is placed in corn syrup, water moves out of the egg, making it smaller.

SAFETY:

Any time you or your students handle chemicals, you should wear safety goggles. An apron is suggested to protect your clothing from vinegar or corn syrup spills. Stress that your students immediately notify you in case of a spill or breakage. Use a pan and brush or paper towels, rather than your bare hands, to pick up broken glass. It is best to use a separate container labeled "broken glass" for the disposal of broken glass to prevent cuts when other people empty the container. Always wash your hands when working with chemicals.

PREPARATION AND HINTS:

1. Students can work alone, or in groups of 2 or 3.
2. You will need 1 raw egg without a shell for each group of students. (It would be wise to prepare several extras. It is unusual to do this activity without breaking any eggs.) You could have your students remove the shells themselves by following the directions in Activity 37 – *Can You Bend a Bone or Bounce an Egg?*. If you are going to remove the shells yourself, place the raw eggs in a glass baking dish,

and cover them with vinegar. You will notice bubbles of carbon dioxide gas on the egg shell almost immediately. The gas is formed as calcium carbonate in the egg shell reacts with the vinegar. The other product of this reaction is calcium acetate, which cannot be seen because it dissolves in the vinegar. The shell will become softer as the calcium carbonate is changed into calcium acetate. If you rub the eggs gently with paper towels every hour, you can remove the shells more quickly. (It takes about 48 hours to remove the shell. It is sometimes easier to leave the eggs over a weekend.) As the shell is removed, the membrane of the egg will be exposed. Discard the vinegar when the shells have been removed. The egg membrane will rupture if poked with a sharp object or squeezed too hard. If you remove the shells at home, the eggs can be transported to school in a covered container. Add some water to the container with the eggs to prevent them from drying out and sticking together.

3. Half of the eggs will be placed in corn syrup. To determine how much corn syrup will be needed, place an egg (with a shell still on) in one of the containers that will be used for the activity, and cover the egg with water. Pour the water into a measuring cup to determine about how much liquid (be it corn syrup or water) will be needed. Label each container of corn syrup with the words "corn syrup".

4. Set up the supply table. Place one "shell–less" egg in each container. Use containers that are wide enough to reach in and pick up an egg, and deep enough so that the egg can be entirely covered with water or syrup. Place labeling supplies (crayons, or markers and masking tape) next to the containers. Put pieces of string that are about 12 inches (30 cm) long in a container of water. (Wet strings are less likely to break the shell membrane.) Place rulers next to the string. It is best to have 1 string and 1 ruler for each egg, but these materials can be shared if necessary. Place the corn syrup on the table. The more organized your supply table is at the beginning of the activity, the more likely it is that supplies will be returned neatly.

5. Place the balance(s) in a location that is easily accessible to all students, and zero the balance(s) (make sure that each balance reads zero when nothing is on the pan, or that both pans are even with each other). Your students should not put their eggs directly on the balance, because if the membrane breaks, the balance could be damaged. The eggs should be placed in a container, and both the container and egg weighed. If possible, have your students weigh the empty container, and subtract the mass of the empty container from that of the egg and container to determine the mass of the egg alone.

6. Provide a waste container for the eggs and the corn syrup. At the end of the activity, you can seal the container before you discard it.

7. Provide paper towels and soap for cleaning the containers, work surfaces, and hands.

8. Provide a garbage can for paper disposal.

9. Discuss the procedure with your students. Stress that the eggs are fragile. The eggs should not be squeezed by their fingers, or by the string as they measure the circumference. The best way to pick the eggs up is to reach under the egg and cradle it in your hand. If you have your student cover their work surface with newspaper or paper towels, clean up will be easier in case an egg breaks (it is inevitable!).

8 Fundamentals of Science Activity Series

TEACHER'S GUIDE
WHAT IS INSIDE AN EGG SHELL?
Parts Of A Raw Egg

GOAL:

1. To observe and identify the parts of a raw egg.

SAFETY:

Goggles should be worn to protect your eyes, and those of your students, from pieces of egg shell that might come loose when the eggs are cracked. The aprons will protect clothing from spills. Stress that your students immediately notify you in case of a spill or breakage. Use a pan and brush or paper towels, rather than your bare hands, to pick up broken glass. It is best to use a separate container labeled "broken glass" for the disposal of broken glass to prevent cuts when other people empty the container. You should be sure that your students wash their hands at the end of this activity.

PREPARATION AND HINTS:

1. Students can work alone, or in groups of 2 or 3.

2. You will need 1 raw egg and 1 container for each group of students.

3. To save time, or if your students would not be able to crack the eggs open without breaking the yolks, you can crack the eggs into the containers of water for them. If a yolk breaks, discard the egg, rinse the container, and begin with a new egg.

4. Set up the supply table. Place each whole egg in a container, or crack each egg into a container of water. Arrange the containers neatly.

5. Provide a waste container for the disposal of eggs that break and for the eggs at the end of the activity. Seal the container before you discard it.

6. Provide a garbage can for the disposal of paper towels and egg shells. (Disposal is neater if the garbage can is first lined with a paper or plastic bag.)

7. Provide paper towels and soap for cleaning up the containers, work surfaces, and hands.

8. Discuss the procedure with your students.

TEACHER'S GUIDE
HOW MUCH VITAMIN C IS THERE IN YOUR FOOD?
Testing For Vitamin C

GOALS:

1. To use an indicator solution to determine the relative amount of vitamin C in different juices.

2. To observe the color change in the indicator.

3. To record the number of drops of juice that are needed to cause a color change in the indicator.

4. To identify the juice sample that has the most vitamin C.

SAFETY:

Goggles should be worn to protect your eyes, and those of your students, from possible splashes of the materials that you use. The aprons will protect clothing from spills and stains. Caution your students not to taste any of the juices used in this activity. Stress that your students immediately notify you in case of a spill or breakage. Use a pan and brush or paper towels, rather than your bare hands, to pick up broken glass. It is best to use a separate container labeled "broken glass" for the disposal of broken glass to prevent cuts when other people empty the container. You should be sure that your students wash their hands at the end of this activity.

PREPARATION AND HINTS:

1. Students can work alone, or in groups of up to 4. Each person should perform the first step of the procedure, to see what a positive test for vitamin C looks like. Each person in the group can then test a different juice sample. Each group member should copy the data from the other group members to complete their Results and Observation charts. If the group is turning in a group report, one person should be designated as the data collector for the group.

2. To prepare the vitamin C indicator you will need cornstarch and medicinal iodine. A stock solution of starch is prepared first by mixing ½ teaspoon of cornstarch with 1 cup (236 mL) of water. The mixture is heated and stirred until the starch is completely dissolved. The stock solution can be used immediately, or it can be refrigerated for

several days before it is used. To make the vitamin C indicator, place 1 teaspoon (5 mL) of the stock starch solution in a container with 1 cup (236 mL) of water. Add 4 drops of medicinal iodine, and stir. The indicator will turn blue as the iodine reacts with the starch. Each group should receive at least ½ cup (118 mL) of the indicator solution. (To prepare 3 quarts of the indicator, mix 4 tablespoons (120 mL) of the stock starch solution with 12 cups (2.8 liters) of water and ¾ teaspoons (3.75 mL) iodine.) Place the vitamin C indicator in labeled containers for each group.

3. Prepare the vitamin C sample by dissolving a vitamin C (250 mg) tablet in 1 cup of water. If you prepare a quart of vitamin C sample, you can place a small amount in a labeled container for each group.

4. Each student will need an eye dropper, and a tablespoon (or 10 mL graduated cylinder). In addition, they will each need a spoon to stir with. Plastic spoons can be used for measuring and stirring.

5. Fresh samples of tomato juice, orange juice, lemon juice, and sauerkraut juice (squeezed from a newly opened can of sauerkraut) should be obtained. The amount of vitamin C in each of these samples decreases rapidly, so keep the samples unopened, or covered and refrigerated, until they are used in class. Pretest the juices to see how many drops of each are required to decolorize the indicator. If the color change occurs within a few drops, you can dilute each of the juices with ¼ cup of water. It should take 5–10 drops of orange juice to cause the color change.

6. Set up the supply table. Place the containers labeled "vitamin C indicator" in one area. In a different area of the supply table, place the juice samples and the containers labeled "vitamin C sample". Arrange the containers neatly. Place the containers (small cups or beakers) on the supply table. Place the tablespoons in a container labeled "tablespoons". Place the spoons to be used for stirring in a container labeled "spoons for stirring". Place the eye droppers on the supply table. The more organized your supply table is at the beginning of the activity, the more likely it is that the supplies will be returned neatly.

7. Provide paper towels and soap for cleaning up the containers, work surfaces, and hands.

8. Discuss the procedure with your students.

9. Discard the vitamin C samples and the indicator solution when you are done with this activity.

TEACHER'S GUIDE
WHERE'S THE FAT?
Testing For Fat In Food

GOALS:

1. To observe the results of the brown paper test for fats.
2. To predict if bacon contains fat.
3. To identify foods that contain fat.

SAFETY:

Use aprons to protect clothing from spills and stains. **Caution your students not to taste any of the foods used in this activity.** Stress that your students immediately notify you in case of a spill or breakage. Use a pan and brush or paper towels, rather than your bare hands, to pick up broken glass. It is best to use a separate container labeled "broken glass" for the disposal of broken glass to prevent cuts when other people empty the container. You should be sure that your students wash their hands at the end of this activity.

PREPARATION AND HINTS:

1. Students can work alone, or in groups of up to 5. Each person should test oil, and at least one other food sample. Each group member should copy the data from the other group members to complete their Results and Observation charts, or one person should be designated as the data collector for the group.

2. Cut brown paper bags into squares that measure approximately 2 inches (5 cm) on each side. You will need at least 10 squares for each group of students. If you put the squares in a flat box, it will be easier for your students to count out the number of squares that they require.

3. Place the samples of butter, honey, oil, peanut butter, and water in small labeled containers. Put a spoon in each container if you don't want your students to use their fingers to remove the samples.

4. Cut the apple and potato into small pieces and put them in labeled containers with enough water to cover the pieces. A ¼ cup of lemon

juice added to the water will prevent the apples from discoloring. Drain the liquid from each container just before your class begins this activity.

5. Cook the pasta and put it in a labeled container with enough water to prevent it from sticking. Each student that tests the pasta will use 1 piece of pasta.

6. Obtain raw ground beef as the meat sample, and place it in a labeled container. Each student that tests the meat will use less than a teaspoon of the ground beef.

7. Cut the bread into small squares and put them in a labeled container.

8. Set up the supply table. Arrange the labeled containers neatly on the table. Place the brown paper squares on the table. The more organized your supply table is at the beginning of the activity, the more likely it is that the supplies will be returned neatly.

9. Provide paper towels and soap for cleaning up the work surfaces and hands.

10. Discuss the procedure with your students.

11. Discard the food samples when you are done with this activity.

TEACHER'S GUIDE
WHERE'S THE STARCH?
A Test For Starch Using Iodine

GOALS:

1. To observe the results of the iodine test for starch
2. To identify foods that contain starch.
3. To apply the observations of positive and negative iodine test results to the results of the iodine test on paper.
4. To infer that paper contains starch because it turns bluish–black when tested with iodine.

SAFETY:

Goggles should be worn to protect your eyes, and those of your students. The aprons will protect clothing from spills and stains. **Caution your students that iodine indicator is a poison, and that they should not taste any of the foods used in this activity.** Stress that your students immediately notify you in case of a spill or breakage. Use a pan and brush or paper towels, rather than your bare hands, to pick up broken glass. It is best to use a separate container labeled "broken glass" for the disposal of broken glass to prevent cuts when other people empty the container. You should be sure that your students wash their hands at the end of this activity.

PREPARATION AND HINTS:

1. Students can work alone, or in groups of up to 5. Each person should test the potato, and at least one other food sample. Each group member should copy the data from the other group members to complete their Results and Observation charts, or one person should be designated as the data collector for the group.

2. Prepare the iodine indicator solution by mixing 1 teaspoon of medicinal iodine with a cup of water. Put the indicator in small, labeled containers. (Old squeeze bottles from contact lens solution work well.) If you do not have squeeze bottles, place an eye dropper into each container.

3. Place the samples of butter and peanut butter in small labeled containers. Put a spoon in each container if you don't want your students to use their fingers to remove the samples.

4. Cut the apple and potato into small pieces and put them in labeled containers with enough water to cover the pieces. A ¼ cup of lemon juice added to the water will prevent the apples from discoloring. Drain the liquid from each container just before your class begins this activity.

5. Cook the pasta and put it in a labeled container with enough water to prevent it from sticking. Each student that tests the pasta will need 1 piece of pasta.

6. Obtain raw ground beef as the meat sample, and place it in a labeled container. Each student that tests the meat will use less than a teaspoon of the ground beef.

7. Cut the crusts from the bread (white bread is best), and discard them. Cut the remaining bread into small squares and put them in a labeled container.

8. Break the crackers (saltines work well) into pieces and put them in a labeled container.

9. Place the sugar cubes in a labeled container.

10. Set up the supply table. Arrange the labeled containers neatly on the table. Place the containers of iodine solution on the table. The more organized your supply table is at the beginning of the activity, the more likely it is that the supplies will be returned neatly.

11. Provide paper towels and soap for cleaning up the work surfaces and hands.

12. Discuss the procedure with your students. Caution them that the iodine can stain their skin and clothing. The stain will eventually wear off their skin, but it will remain on their clothing. Remind them that iodine is a poison, and that they should not eat any of these food samples.

13. Discard the food samples when you are done with this activity.

14. To conclude this activity, write a secret message to your students and make it appear magically by putting it in a pan and covering it with iodine indicator. Prepare the "ink" by carefully mixing 1 teaspoon of cornstarch in ½ cup of water. Find some paper that was not made with starch. (Test the paper with iodine indicator. The iodine indicator spot will remain reddish–brown on paper that is not made with

starch. Newspaper that doesn't have print on it can generally be used.) Write your message on the paper using your "ink" and a small brush or a cotton swab, and allow it to dry. Show your class the blank paper, and then put it in a pan and add the iodine solution. Challenge them to figure out how you made the message magically appear.

15. A second challenge. Bring in a piece of toast and have your students predict the results of the iodine test. The untoasted, white, inner portion will turn bluish–black, while the brown, toasted outer surfaces and crusts will turn red. Toasting actually breaks down the starch in the outer part of the bread into a smaller molecule called dextrin. The red color is produced when the iodine reacts with the dextrin.

16 Fundamentals of Science Activity Series

TEACHER'S GUIDE
WHAT MAKES ORANGE KOOL AID ORANGE?

Chromatography To Detect Artificial Coloring In Foods

GOALS:

1. To use the process of chromatography to separate the pigments in orange Kool Aid.

2. To prepare a paper chromatogram.

3. To observe the finished chromatogram of Kool Aid, and compare it to the original appearance.

4. To conclude that Orange Kool Aid contains more than one artificial color.

SAFETY:

Goggles should be worn to protect your eyes, and those of your students. The aprons will protect clothing from spills and stains. Stress that your students immediately notify you in case of a spill or breakage. Use a pan and brush or paper towels, rather than your bare hands, to pick up broken glass. It is best to use a separate container labeled "broken glass" for the disposal of broken glass to prevent cuts when other people empty the container. You should be sure that your students wash their hands at the end of this activity.

PREPARATION AND HINTS:

1. To simplify this activity for younger students, cut strips of paper towel that are 1 inch (2.5 cm) wide. Draw a line of water soluble (washable) marker across the strip, 2 inches (5 cm) from the bottom. Arrange the strip in a shallow dish of water so that the part of the paper below the marker line is in the water, but the marker line and the rest of the strip stay dry. You will see that the strip becomes wet, and as the water moves up to the top of the paper, the different colors of ink in the marker dissolve in the water and are carried up also. One or more bands of color appear as the ink molecules with the least attraction for water are left behind as the water continues to rise up the paper. Different color markers will produce different color bands. Black markers usually produce several bands of color. Water color

paints can be mixed together and painted on the towel strip in the same way. After the paint dries, place the bottom of the paper in water. The mixed colors will be separated into different color bands on the paper strip. Permanent markers can also be used, but the solvent in the dish should be alcohol (rubbing alcohol).

2. Students can work alone, or in pairs.

3. Prepare the Orange Kool Aid by mixing a package of orange Kool Aid in ¼ cup (60 mL) of water. Stir the concentrated mixture to dissolve the Kool Aid.

4. Place the concentrated Kool Aid mixture in small containers labeled "Orange Kool Aid". Put a toothpick or narrow brush in each container. (One container per student is ideal, but if your students will have to share the Kool Aid, place several toothpicks or brushes in each container.) If the mixture dries out, add several drops of water, and stir to dissolve the Kool Aid.

5. Prepare the solvent by mixing equal amounts of water and rubbing alcohol. Put the solvent in several containers that can be capped or covered with plastic wrap or foil. Label each container "Solvent".

6. Collect plastic lids that can be used for tracing circles on the paper towels. (If you don't want your students to handle scissors, or if you want to save time in class, you can trace the circles, cut the circles out, and cut the "tails" yourself.) (If you stack the towels, you can cut several at a time.)

7. Collect the containers (small cups or beakers work well) that you will use for the activity. If you don't have graduated cylinders to measure 25 mL of solvent, you can have your students place 2 – 3 tablespoons of solvent in the container. To save time, you can add the solvent to each container yourself. Do this just before you begin this activity, as the alcohol in the solvent will evaporate more rapidly than the water, unless you cover each container.

8. Set up the supply table. Arrange the labeled containers of Kool Aid and solvent neatly on the table. Place the paper towels, plastic lids, and scissors (or the precut circles) on the table. Organize the containers that you will use, and the graduated cylinders or measuring spoons on the table. The more organized your supply table is at the beginning of the activity, the more likely it is that the supplies will be returned neatly.

9. Provide paper towels and soap for cleaning up the work surfaces and hands.

10. Discuss the procedure with your students. Caution them that the concentrated Kool Aid can stain their skin and clothing. The stain will eventually wear off their skin, but it will probably remain in their clothing. Remind them that the level of the solvent in the container should be <u>below</u> the line of Kool Aid on their paper "tail", and that their line of Kool Aid should be dry before they place the tail in the solvent.

11. Discard the solvent when you are done with this activity.

12. Challenge your students to explain how a leaf that is green all summer can become yellow, orange or red in the fall. (The green color of the leaves is so dark that it hides the other colors. In the fall, the decreased number of daylight hours cause the green pigment to decompose. We can then see the other color(s). The hidden colors in green leaves can be separated by paper chromatograpy if special solvents are used.)

TEACHER'S GUIDE
CABBAGE MAGIC: HOW TO CHANGE RED CABBAGE JUICE TO A PINK OR GREEN COLOR

Red Cabbage Indicator For Acids And Bases

GOALS:

1. To extract juice from red cabbage to be used as an indicator.

2. To observe and record the color changes that occur when red cabbage indicator is combined with an acid, and then with a base.

3. To test different samples with red cabbage indicator, and to determine which samples are acids and which are bases by observing the color changes in the red cabbage indicator.

4. To conclude that when the indicator turns pink, the sample is an acid, and that when the indicator turns green, the sample is a base.

SAFETY:

Goggles should be worn to protect your eyes, and those of your students. Aprons will protect clothing from spills and stains. If your students are going to cut and grate the red cabbage to make the extract, caution them that the knives and graters are sharp, and that they can cut themselves if they are careless. Also be sure that they are well versed in the safe use of a stove. Train your students to immediately notify you in case of a spill or breakage. Use a pan and brush or paper towels, rather than your bare hands, to pick up broken glass. It is best to use a separate container labeled "broken glass" for the disposal of broken glass to prevent cuts when other people empty the container. You should be sure that your students wash their hands at the end of this activity.

PREPARATION AND HINTS:

1. To simplify this activity for younger students, prepare the red cabbage extract at home. Dilute each cup of extract with 3 cups of water to produce the red cabbage indicator. Refrigerate the indicator until you use it. If the red cabbage indicator is placed in small, labeled squeeze bottles, 1 tablespoon (15 mL) can easily be measured and placed into the test container. If the samples are also placed in labeled squeeze bottles, they can easily be measured and added to the

indicator. (Students can count the drops as they add the sample directly to the indicator, or squeeze the sample into quarter teaspoons and count the number of quarter teaspoons that they add to the indicator, or simply observe and record the color change in the indicator as they add the sample without measuring it.)

2. Students can work alone, or in groups of up to 4. Each student in the group should test lemon juice, and at least one other sample.

3. Place the samples of lemon juice, orange juice, sauerkraut juice (from a can of sauerkraut), soda, and vinegar in small labeled containers. If the containers are not squeeze bottles, put a ¼ teaspoon or an eye dropper (15 drops are approximately ¼ teaspoon) in each container. It is best if you can prepare a set of these containers for each group. Your students can share the samples if they remember to return each sample when they are done using it, and if they don't all use the same sample simultaneously.

4. Place the ammonia in a secure location to be used when you demonstrate the effect of a base on red cabbage indicator.

5. Set up the supply table. Arrange the labeled containers of red cabbage indicator in one area, and the samples to be tested in another area. Place the containers, measuring spoons (tablespoons for the indicator, ¼ teaspoons or eye droppers for the samples) and stirring spoons neatly on the table.

6. Provide paper towels and soap for cleaning up the work surfaces, containers, spoons, and hands.

7. Discuss the procedure with your students. Caution them that the red cabbage indicator might stain their skin and clothing. The stain will eventually wear off their skin, but it may remain in their clothing.

8. Discard the red cabbage indicator and samples when you have completed this activity. If you intend to do Activity 9 – *How Acidic is It?*, within a few days, you can refrigerate the samples and indicator.

9. Challenge your students either before you begin this activity, or after you complete the activity, to explain how you can pour a purple liquid into a glass that contains a clear liquid and magically have the final color be pink. (The clear liquid is vinegar, and the purple liquid is red cabbage indicator.) How can you pour a purple liquid into a glass that has a clear liquid, and magically have the final color be green? (The clear liquid is ammonia, and the purple liquid is red cabbage indicator).

Basic Biology Science Activities 21

TEACHER'S GUIDE
HOW ACIDIC IS IT?

Using Red Cabbage Indicator To Determine Which Is The Most Acidic Substance

GOALS:

1. To extract juice from red cabbage to be used as an indicator.

2. To observe and record the color changes that occur when red cabbage indicator is combined with an acid and a base.

3. To test different substances with red cabbage indicator.

4. To identify the substance that is the most acidic by quantitative means.

5. To observe the effect of adding a base to an acid solution.

6. To conclude that an acid can be neutralized by the addition of a base.

SAFETY:

Goggles should be worn to protect your eyes, and those of your students. Aprons will protect clothing from spills and stains. If your students are going to cut and grate the red cabbage to make the extract, caution them that the knives and graters are sharp, and that they can cut themselves if they are careless. Also be sure that they are well versed in the safe use of a stove. Train your students to immediately notify you in case of a spill or breakage. Use a pan and brush or paper towels, rather than your bare hands, to pick up broken glass. It is best to use a separate container labeled "broken glass" for the disposal of broken glass to prevent cuts when other people empty the container. You should be sure that your students wash their hands at the end of this activity.

PREPARATION AND HINTS:

1. To save time, you can prepare the red cabbage indicator at home according to the directions in the procedure. Refrigerate the indicator until you use it. If the red cabbage indicator is placed in small, labeled squeeze bottles, 1 tablespoon (15 mL) can easily be measured and placed into the test container.

2. Students can work alone, or in groups of up to 4. Each student should test lemon juice and at least one other sample.

3. Place the samples of lemon juice, orange juice, rain water, sauerkraut juice (from a can of sauerkraut), soda, tap water, and vinegar in small labeled containers. If you place the samples in labeled squeeze bottles, it is easy to count the number of drops that are added to the indicator as the container is squeezed. If you do not have a sufficient number of squeeze bottles, place an eye dropper in each container. Traffic flow around the supply table will be reduced if you can prepare a set of these containers for each group that is to do this activity. If this is not possible, your students can share the samples if they remember to return each sample when they are done using it, and if they don't all use the same sample simultaneously.

4. Place the ammonia in a secure location. You will need it for the demonstration of the number of drops of ammonia needed to cause a color change in the red cabbage indicator. Your students will need to use it under your direct supervision to see how many drops of ammonia it takes to neutralize the acid in step 10. (You can demonstrate this step if you don't want your students to handle the ammonia.)

5. Set up the supply table. Arrange the labeled containers of red cabbage indicator in one area, and the samples to be tested in another area. Place the containers, tablespoons to measure the indicator, stirring spoons, and eye droppers neatly on the table.

6. Provide paper towels and soap for cleaning up the work surfaces, containers, spoons, and hands.

7. Discuss the procedure with your students. Caution them that the red cabbage indicator might stain their skin and clothing. The stain will eventually wear off their skin, but it may remain in their clothing.

8. Discard the red cabbage indicator and samples when you have completed this activity.

9. Challenge your students either before you begin this activity, or after you complete the activity, to explain how you can pour a clear liquid into a purple liquid and have the mixture turn pink, and then add another clear liquid and have the mixture turn green. (The first clear liquid is vinegar, the purple liquid is red cabbage indicator, and the second clear liquid is ammonia. You are neutralizing the acid (vinegar) with the base (ammonia).

Test this before you challenge your class to be sure that you have the correct amounts of vinegar and ammonia. Usually ½ cup of vinegar added to several tablespoons of red cabbage indicator will be neutralized by ½ cup of ammonia.)

24 Fundamentals of Science Activity Series

TEACHER'S GUIDE
DO YOU EAT FOOD THAT CONTAINS BACTERIA?

Detection Of Bacteria Using Methylene Blue

GOALS:

1. To observe that the color of methylene blue solution is changed by the addition of certain foods

2. To identify the control as the "methylene blue" that has nothing added to it.

3. To conclude that certain foods that we eat contain living bacteria.

SAFETY:

Goggles should be worn to protect your eyes, and those of your students. Aprons will protect clothing from spills and stains. Train your students to immediately notify you in case of a spill or breakage. Use a pan and brush or paper towels, rather than your bare hands, to pick up broken glass. It is best to use a separate container labeled "broken glass" for the disposal of broken glass to prevent cuts when other people empty the container. You should be sure that your students do not eat any of the samples used in this activity, and that they wash their hands at the end of this activity.

PREPARATION AND HINTS:

1. Students can work alone, or in groups of up to 4. Each student in the group should test at least one sample.

2. Place methylene blue solution in several small containers. Label each container with the words "Methylene Blue Solution". The solution can be dispensed with eye droppers. You can also use small plastic squeeze containers. (Small contact lens squeeze bottles work well. These containers should not be used for any other purpose after they are used for methylene blue. They can be capped and stored for other activities that use methylene blue solution.)

3. Place the oil in several containers. Plastic containers that have small openings (like contact lens wetting solution containers) reduce the number of spills, are easy to use, and can be capped and stored.

4. Place the samples of milk, yogurt, buttermilk and cottage cheese in small, labeled containers. Place several teaspoons in each container to be used to measure each sample. Cover the containers and refrigerate them until needed.

5. Set up the supply table. Arrange the labeled containers of methylene blue solution in one area, and the samples to be tested in another area. Place the containers (cups, jars or test tubes in a test tube rack), and spoons to be used for stirring neatly on the table.

6. Provide paper towels and soap for cleaning up the work surfaces, containers, spoons, and hands.

7. Discuss the procedure with your students. Caution them not to taste any of the food samples used in this activity. Make them aware of the fact that the methylene blue solution might stain their skin and clothing. The stain will eventually wear off their skin, but it may remain in their clothing.

8. Discard the food samples when you have completed this activity.

9. When you have completed this activity, challenge your students to explain the results of the demonstration that is in front of them. Tell your students that you prepared three containers, as they did, and added yogurt to two of the containers. The third container had no yogurt added, thus serving as the control. They can observe that the control stayed blue, and that one of the yogurt containers is no longer a blue color. They will also see that the other container of yogurt has a layer of blue on the surface. Ask them to explain why one container has a layer of blue on the surface.

To prepare this demonstration, one day in advance obtain three transparent containers. Add water to half fill each container, and add 10 drops of methylene blue solution to each. Add a teaspoon of yogurt to two of the containers and stir. Pour oil to a depth of 0.5 cm on the surface of one of the containers that has yogurt in it, and on the surface of the container that has no yogurt (this container is your control). The next day, bring out the three containers.

The explanation for the container that has blue on the surface is that oxygen in the air reacted with the methylene blue in the yogurt container that was **not** covered with a layer of oil. Although the bacteria in the yogurt changed the methylene blue to methylene white, the oxygen in the air changed the methylene white on the surface back to methylene blue, producing a surface layer of blue.

26 Fundamentals of Science Activity Series

TEACHER'S GUIDE
WHAT DO YOU GET WHEN YOU MAKE THIN SLICES OF CARROTS AND CELERY?

How To Make A Cross Section

GOALS:

1. To identify and remove a vein from a stalk of celery.
2. To draw an accurate sketch of celery and carrot cross sections.
3. To correctly identify the pith and veins in a celery cross section.
4. To predict what a cross section of a carrot will look like before a carrot is sliced.

SAFETY:

Aprons will protect clothing from spills and stains. Be sure that your students are cautioned that they could cut themselves if they use the knives carelessly. Train your students to immediately notify you in case of a spill or breakage. You should be sure that your students do not eat any of the vegetables used in this activity, and that they wash their hands at the end of this activity.

PREPARATION AND HINTS:

1. An effective way to start this activity is to bring in an unusual fruit or vegetable that you have sliced prior to class. Show your students the slices, and have them guess (predict) what the original fruit or vegetable looked like. Bring an unsliced piece of the fruit or vegetable to show them the accuracy of their predictions. (Star fruit makes beautiful yellow star shaped slices.)

2. If you don't want your students to handle knives to slice the celery or carrot, you can slice the vegetables for them. You can also make the cross sections for your students to save time during class. Be sure to have them predict what the carrot cross sections will look like before they actually see the slices. You can also save time in class if you cut off the bottom of several stalks of celery and place the cut ends in red ink or red food coloring and water. (Do this several hours or the day before you begin the activity with your students.) The water will move

up the veins, staining them red. The stalks can then be used for procedures 4, 7, and 8.

3. Students can work alone, or in pairs.

4. Prepare the supply table. Place the carrots and celery stalks on the table. Even if you have sliced the celery for your students, they will each need a stalk to try to remove a vein. If you have sliced the carrots, do not put them on the table at this time. (Wait until they have made their predictions.) If your students are to slice the vegetables themselves, place the knives in a "cutting area" so that your students can use the knives while you are watching them. If your students are going to do procedure 4, 7, and 8, place food coloring (or ink) on the table, as well as containers that will support the celery stalks in the colored water.

5. Provide paper towels and soap for cleaning up the work surfaces, containers, and hands.

6. Discuss the procedure with your students. Caution them not to taste the vegetables used in this activity. Make them aware of the fact that the food coloring used in procedure 4 might stain their skin and clothing. The stain will eventually wear off their skin, but it may remain in their clothing.

7. As your students observe their carrot cross sections, they may ask how to label the parts of their sketches. This is not called for in the lab activity, but the innermost part of the circle is composed of xylem cells, that are surrounded by a thin layer of phloem cells. The outer ring of the carrot slice is called the cortex.

8. Discard the vegetables when you have completed this activity. You can save the celery if you are going to do Activity 12 – *How Does Water Move Up A Plant Stem?*.

9. When you have completed this activity, challenge your students to apply their knowledge of cross sections. Ask them to take 2 or 3 different colors of clay, and shape the colored clay in such a way that when you slice the clay, you get circles of clay that have the appearance of a bulls eye target. (You could even specify the sequence of colors in the "bulls eye".) You could do a similar activity if your school has cooking facilities by preparing cookie dough, and using food coloring to dye portions of the dough different colors. Have your students combine the pieces of different colored dough, and predict what the cookie cross sections will look like before they are sliced. After baking, the students can eat their cookie cross sections.

28 Fundamentals of Science Activity Series

TEACHER'S GUIDE
HOW DOES WATER MOVE UP A PLANT STEM?
Transport In A Plant Stem

GOALS:

1. To identify and remove a vein from a stalk of celery.

2. To draw accurate sketches that document the changes in the appearance of the celery stalk as it absorbs colored water.

3. To identify the part of a plant stem that carries water to the leaves.

SAFETY:

Aprons will protect clothing from spills and stains. Be sure that your students are cautioned that they could cut themselves if they use the knives carelessly. Train your students to immediately notify you in case of a spill or breakage. You should be sure that your students do not eat any of the samples used in this activity, and that they wash their hands at the end of this activity.

PREPARATION AND HINTS:

1. If your students have completed Activity 11 – *What Do You Get When You Make Thin Slices of Carrots and Celery?*, they can skip procedures 1 – 3, and the first two items in the Results and Observations section.

2. To save time, you could do procedure steps 4 – 11 for your students. Several stalks of celery can be placed in each container of colored water.

3. Students can work alone, or in pairs.

4. Prepare the supply table. Place the celery stalks on the table. If your students are to slice the celery stalks themselves, place the knives in a "cutting area" so that they can use the knives while you are watching them. Place containers that can support the celery stalks in the colored water on the table. Place several jars of food coloring on the table. (You can prepare the colored water in advance, and pour it into their containers if you don't want them to get red food coloring on themselves.) If available, crayons or colored pencils (to make

accurate sketches) should be provided. The more organized your supply table is at the beginning of the activity, the more likely it is that your students return materials neatly.

5. Provide paper towels and soap for cleaning up the work surfaces, containers, and hands.

6. Discuss the procedure with your students. Caution them not to taste the celery stalks used in this activity. Make them aware of the fact that the food coloring might stain their skin and clothing. The stain will eventually wear off their skin, but it may remain in their clothing.

7. Discard the celery when you have completed this activity.

8. When you have completed this activity, challenge your students to explain how you could have a flower that is half red and half blue. (The day before, slit several inches of the stem of a white carnation. Gently separate the halves of the stem, and place one half in a container that has red food coloring added to the water. Place the other half of the stem in a container that has blue food coloring added to the water. The darker the colors in the water are initially, the more obvious the differences will be in the flower the next day. Before you show your "rare and unusual" flower to your students, take the stem out of the containers of food coloring, and place it in a vase or wrap it in tissue paper.)

9. Have a contest to see who can best represent a season or holiday with a colored flower. (A Fourth of July carnation can be made by dividing the stem of a white carnation into thirds, and then placing one part of the stem in red colored water, one part in blue colored water, and one part in plain water.)

TEACHER'S GUIDE
THE END OF THE LINE?
Transpiration In Plants

GOALS:

1. To define transpiration.
2. To observe that water is released by a plant.

SAFETY:

Be sure that your students are cautioned that they could cut themselves if they use the knives carelessly. Train your students to immediately notify you in case of a spill or breakage. Be sure that your students wash their hands at the end of this activity.

PREPARATION AND HINTS:

1. To save time, or if you don't want your students to handle the knives or razor blades, you could do procedure 2 for them, and place the cut ends of the stems in a container of water.

2. Students can work alone, or in pairs.

3. Collect jars that can be used to cover the plants that have been placed in cups. Gallon mayonnaise jars or other transparent, industrial sized containers can be obtained from the school cafeteria, a restaurant or from a delicatessen. Two or three liter plastic soda bottles can also be used if the narrow part of the neck is cut off with a knife. (If you use plastic soda bottles, save both portions to be used for Activity 22 – *Can You Grow A Plant Without Soil?*) If it is difficult to obtain these large containers, an empty aquarium can be inverted over all of the cups in procedure 5. A second empty aquarium can be used for the control cups in procedure 6.

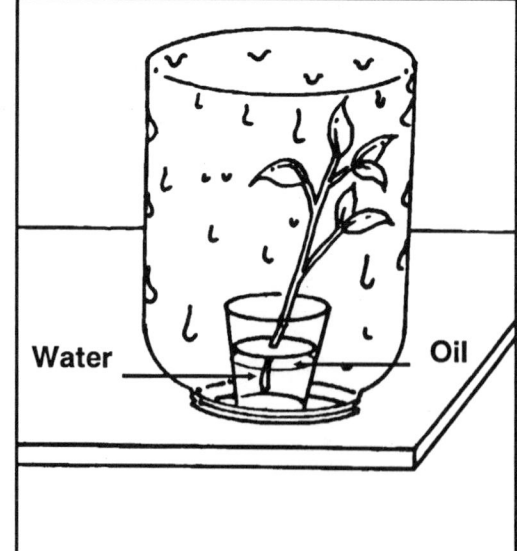

4. Place the oil in several squeeze bottles. (Contact lens wetting solution bottles work well.) It is easy to use squeeze bottles to put oil on the surface of the water in procedures 3 and 6 because the bottles won't break, and they are less likely to spill than open containers of oil. Covering the surface of the water with oil assures that any water that later condenses on the inner surface of the jar is the result of water that has been lost from the plant by transpiration rather than water that has evaporated from the cup.

5. You will get faster results if your students leave their covered plants in a warm, sunny location. They can even seal the jar to the counter top with pieces of clay to prevent the water that is lost by transpiration from escaping into the atmosphere through gaps between the surface of the counter, and the jar that covers the plant. (This is an important step if you are using soda bottles instead of the glass jars, because the cut edges of the bottles are probably irregular and might allow the water to escape into the atmosphere before it has a chance to condense on the inside of the container).

6. Prepare the supply table. Place the plants on the table. If your students are to slice the plants themselves, place the knives or razor blades in a "cutting area" so that they can use the sharp instruments while you are watching them. Place containers that can support a plant stem in water on the table. Place the oil bottles on the table. The more organized your supply table is at the beginning of the activity, the more likely it is that your students return materials neatly.

7. Provide a waste can for the oil and water. The oil should not be discarded in the sink. Seal the container before you dispose of it in the garbage.

8. Provide paper towels and soap for cleaning up the work surfaces, containers, and hands.

9. Discuss the procedure with your students. Stress that they should not lift the jar off the plant once they put it over the plant in procedure 5. (This would allow any moisture that has been lost by transpiration to escape into the atmosphere, rather than to build up and finally condense on the inside of the jar.)

10. At the end of the activity, you can save the cut pieces of the plants by rooting them. (See Activity 23 – *Can You Grow A New Plant Without Using A Seed?* for the procedure. Your students could even do this at home.)

32 Fundamentals of Science Activity Series

TEACHER'S GUIDE
SEED PACKAGES
Parts Of A Bean Seed

GOALS:

1. To identify the seed coat, cotyledon, embryo, epicotyl, and hypocotyl of a seed.

2. To describe the function of the seed coat, cotyledon, embryo, epicotyl, and hypocotyl of a seed

3. To sketch the embryo, and correctly label the epicotyl and hypocotyl.

SAFETY:

You should be sure that your students do not eat the seeds used in this activity, and that they wash their hands at the end of this activity.

PREPARATION AND HINTS:

1. Soak large, dry, bean seeds in water for 24 hours before you begin the activity. (It is easier to see the parts of the seeds if you use large beans.) To do this, place the dry beans in a container, and add enough water to cover the beans. The next day, pour off the water before you put the container on the supply table.

2. Students can work alone, or in pairs.

3. Set up the supply table by placing the container of beans on the table. Place hand lenses or magnifying glasses near the seeds. Your students will also need red and blue crayons, or colored pencils. If they are going to save the Experiment Page, they will need tape to attach the seed parts to the paper.

4. Provide paper towels and soap for cleaning up the work surfaces and hands.

5. Discuss the procedure with your students. Remind your students that the seed is very small, and must be handled gently or they may damage a part of the seed before they get a chance to observe it.

TEACHER'S GUIDE
WHAT PART OF A SEED STORES FOOD FOR A BABY PLANT?

Starch In A Bean Seed

GOALS:

1. To observe the effect of iodine on the seed coat, cotyledon, and embryo of a seed.

2. To identify the part of the seed that contains starch, based on the results of the iodine test.

3. To conclude that the cotyledon stores starch.

SAFETY:

Caution your students that the iodine is a poison. It can temporarily stain their hands, and permanently stain their clothes. Goggles should be worn to protect their eyes and aprons should be worn to protect their clothing. You should be sure that your students do not eat the seeds used in this activity, and that they wash their hands at the end of this activity.

PREPARATION AND HINTS:

1. Soak large, dry, bean seeds in water for 24 hours before you begin the activity. To do this, place the dry beans in a container, and add enough water to cover the beans. The next day, pour off the water before you put the container on the supply table.

2. Prepare the iodine indicator solution by mixing 1 teaspoon (5 mL) of medicinal iodine with 1 cup (236 mL) of water. Put the indicator in small, labeled containers. (The empty squeeze bottles of contact lens solution work well. They are less likely to spill or break than other containers, and they can be capped for storage.) If you do not have squeeze bottles, put several eye droppers in each container.

3. Students can work alone, or in pairs.

4. Set up the supply table by placing the container of beans on the table. Place hand lenses or magnifying glasses and the containers of iodine solution near the seeds. If your students are going to save the

Experiment Page, they will need tape to attach the seed parts to the paper.

5. Provide paper towels and soap for cleaning up the work surfaces and hands.

6. Discuss the procedure with your students. Remind your students that the seed is very small, and must be handled gently or they may damage a part of the seed before they get a chance to observe it.

7. The cotyledons <u>should</u> test positively for starch. If they do not turn bluish–black when tested with iodine, the surface of the cotyledons may have to be scratched slightly. An opened paper clip can be used to scratch the cotyledons, thus loosening some of the starch granules, which will test positively for starch.

TEACHER'S GUIDE
SEED RACES

Which Seeds Germinate And Grow The Fastest?

GOALS:

1. To observe the process of seed germination.
2. To compare and record the germination rates of several different types of seeds.
3. To compare and record the growth rates of several different types of seeds.
4. To identify at least one of the factors needed for a seed to germinate.

SAFETY:

Caution your students that the seed disinfecting solution is a poison, and that they should wash their hands after they handle the seeds. They should wash until their hands no longer feel slippery. Goggles should be worn to protect eyes from the disinfecting solution, and aprons should be worn to protect clothing. You should be sure that your students do not eat the seeds used in this activity.

PREPARATION AND HINTS:

1. Younger students can do this activity if you do procedure steps 1 – 3 for them. Have them place a radish seed and 1 or 2 other seeds in their bags, and hang the bags on a bulletin board. Their names can be written on the bag before or after they add the seeds, using a permanent marker. Your students can observe their seeds germinate and grow without removing them from the board.

2. Students can work alone, or in pairs.

3. Soak two or three kinds of dry seeds (kidney beans, black beans, white beans, navy beans, mung beans, chick peas, lentils or corn seeds) in separate labeled containers for 24 hours before you begin the activity. To do this, place each type of dried seeds in a separate container, and add enough water to cover the seeds. (Each student that "enters" the seed race will use 3 – 6 different seeds.) Drain the seeds before you disinfect them.

4. In addition to choosing 2 or 3 kinds of seeds from the list above, choose 1 or 2 seeds such as oat, wheat, and radish. These seeds do not have to be soaked for 24 hours before the seed race. The radish usually germinates before all of the other seeds. If your students need quick results, this is one of the seeds that you should use.

5. To prevent the seeds from getting moldy (and being disqualified from the race), immerse them in a seed disinfecting solution for 2 minutes. A seed disinfecting solution can be made by mixing ¼ cup (59 mL) of bleach with 1 quart (944 mL) of water. The easiest way to disinfect the seeds is to place them in a strainer, and then dip the strainer and the seeds into the disinfecting solution for two minutes. A straining bag can be also be made using discarded stockings, or nylon netting. After the seeds have drained, return each type of seed to its own clean, labeled container. The seeds are ready to be used immediately.

6. Obtain ziplock plastic bags (larger than sandwich bag size, if possible), paper towels, staplers (borrow staplers if possible, so there are several staplers available), and extra staples.

7. Prepare a seed bag before the activity begins so that your students can see what their bags should look like. You can even prepare a separate bag for each step of the procedure and tape the bags to the blackboard for reference.

8. Set up the supply table. Place the labeled containers of disinfected seeds in one area, and the ziplock bags, paper towels, staplers and extra staples in another area. If you have enough extra seeds, you can provide dry seeds that have not been disinfected for procedure 4. Permanent markers, crayons or pencils can be used to identify each bag.

9. Provide paper towels and soap for cleaning up the work surfaces and hands.

10. Discuss the procedure with your students. Remind them to wash their hands carefully as soon as they have placed the seeds in their bag, and have sealed it shut. Show them where they are to bring their seed bags when they have finished preparing them. The bags should be hung on a bulletin board, wall, or black board using staples, thumb tacks, or tape.

Basic Biology Science Activities 37

TEACHER'S GUIDE
AT WHAT TEMPERATURE DO SEEDS SPROUT?
Seed Germination vs. Temperature

GOALS:

1. To define germination.

2. To determine which temperature is most suitable for seed germination.

3. To calculate the percentage of seed germination at different temperatures

4. To explain why it is important that seeds germinate only in specific environmental conditions.

SAFETY:

Caution your students that the seed disinfecting solution is a poison, and that they should wash their hands after they handle the seeds. They should wash until their hands no longer feel slippery. Goggles should be worn to protect eyes from the disinfecting solution, and aprons should be worn to protect clothing. You should be sure that your students do not eat the seeds used in this activity.

PREPARATION AND HINTS:

1. Younger students can do this activity if you omit procedure 6, the calculation of percentage germination.

2. This activity should be done in groups of three. Each student in the group will prepare 1 roll of seeds in a bag. Each of the bags will be stored at a different temperature. The group members should exchange the data that they obtain regarding seed germination at their assigned temperature.

3. This activity can be modified to determine just the percentage germination (rather than the percentage germination at different temperatures). Each student prepares 1 bag of seeds rolled in damp towels according to procedure steps 1 – 5. The seed bags should all be kept at room temperature. The data chart can be used if your students leave the columns for "cold" and "warm" blank.

4. If you are using large, dried seeds (kidney beans, black beans, white beans, navy beans, mung beans, chick peas, or corn seeds) they should be soaked for 24 hours before you begin the activity. Place the dried seeds in a container, and add enough water to cover the seeds. (Each student that prepares a bag will use 10 seeds, so don't soak more seeds than are needed.) Drain the seeds before you disinfect them.

5. If you are using radish or mustard seeds, they do not have to be soaked in advance.

6. To prevent the seeds from getting moldy and not germinating, immerse them in a seed disinfecting solution for 2 minutes. A seed disinfecting solution can be made by mixing ¼ cup (59 mL) of bleach with 1 quart (944 mL) of water. The easiest way to disinfect the seeds is to place them in a strainer, and then dip the strainer and the seeds into the disinfecting solution for two minutes. A straining bag can also be made using discarded stockings or nylon netting. After the seeds have drained, place them in a clean, labeled container. The seeds are ready to be used immediately.

7. Obtain ziplock plastic bags (sandwich bag size is adequate), paper towels, permanent markers for labeling, and an incubator.

8. You can make an incubator by putting a light inside a large cardboard box. You can also use an electric heating pad on the lowest setting. Either of these techniques should produce temperatures that are higher than the surrounding room temperature. (If you have a thermometer you can measure the temperature. The closer the temperature is to body temperature, the lower the germination rate will be.)

9. Prepare up a sample seed bag so that your students can see what their bags should look like.

10. Set up the supply table. Place the disinfected seeds in one area, and the ziplock bags, paper towels, and markers in another area.

11. Provide paper towels and soap for cleaning up the work surfaces and hands.

12. Discuss the procedure with your students. Remind them to wash their hands as soon as they have the seeds in the bag, and have sealed it shut. Consider demonstrating the preparation of a "seed roll" before you show your students what the finished seed bag will look like. Show them where to store their seeds (the three different temperature areas) when they have completed their seed bags.

TEACHER'S GUIDE
CAN YOU GROW A PLANT UPSIDE DOWN?
Geotropism

GOALS:

1. To define geotropism.

2. To observe the effect of gravity on the growth of roots and stems

3. To explain why positive and negative geotropism help a plant to survive.

SAFETY:

Caution your students that the seed disinfecting solution is a poison, and that they should wash their hands after they handle the seeds. They should wash until their hands no longer feel slippery. Goggles should be worn to protect eyes from the disinfecting solution, and aprons should be worn to protect clothing. You should be sure that your students do not eat the seeds used in this activity. Straight pins can be dangerous if used carelessly.

PREPARATION AND HINTS:

1. This activity can be done individually, or in groups of up to four.

2. Preparing the bags with the staples and pins is time consuming and can be dangerous for young students. If the seeds are pressed down so that they rest firmly on the first row of staples, and the second row of staples is placed as close to the seeds as possible, the seeds will probably remain in position without using pins to make individual compartments. Another way to save time is to sandwich a sponge and the seeds between squares of glass or clear plastic. This procedure is described in Suggestions For Further Studies, item 3.

3. One type of large, dried seeds (kidney beans, black beans, white beans, navy beans, mung beans, chick peas, or corn seeds) should be soaked for 24 hours before you begin the activity, to hasten germination. To do this, place the dried seeds in a container, and add enough water to cover the seeds. (Each student that prepares a bag will use 10 seeds, so don't soak more seeds than will be needed.) Drain the seeds before you disinfect them.

4. To prevent the seeds from getting moldy and not germinating, immerse them in a seed disinfecting solution for 2 minutes. A seed disinfecting solution can be made by mixing ¼ cup (59 mL) of bleach with 1 quart (944 mL) of water. The easiest way to disinfect the seeds is to place them in a strainer, and then dip the strainer and the seeds into the disinfecting solution for two minutes. A straining bag can be also be made using discarded stockings, or nylon netting. After the seeds have drained, place them in a clean, labeled container. The seeds are ready to be used immediately.

5. Obtain ziplock plastic bags (sandwich bag size is adequate), paper towels, permanent markers for labeling, several staplers, extra staples, and straight pins.

6. Prepare a sample of the bag so that your students can see what their bags should look like. Tape it to the black board for reference.

7. Set up the supply table. Place the disinfected seeds in one area, and the ziplock bags, paper towels, staplers, pins and markers in another area.

8. Provide paper towels and soap for cleaning the work surfaces and hands.

9. Discuss the procedure with your students. Remind them to wash their hands as soon as the seeds are in the bag, and the bag is sealed shut. Show them where to hang their seed bag when it has been completed.

TEACHER'S GUIDE
CAN THE ROOTS FIND THE WATER?
Hydrotropism

GOALS:

1. To define hydrotropism.
2. To observe the growth of roots towards a water source.
3. To explain why hydrotropism is important for plant survival.

SAFETY:

Caution your students that the seed disinfecting solution is a poison, and that they should wash their hands after they handle the seeds. They should wash their hands until they no longer feel slippery. Goggles should be worn to protect eyes from the disinfecting solution, and aprons should be worn to protect clothing. You should be sure that your students do not eat the seeds used in this activity. Straight pins and scissors can be dangerous if used carelessly.

PREPARATION AND HINTS:

1. This activity can be done individually, or in groups of up to four.

2. Obtain corrugated cardboard, and cut it into rectangular pieces that measure approximately 5 inches x 10 inches (12 cm x 24 cm). The cardboard will fold in half more easily if you score each piece first.

3. Corn seeds should be soaked for 24 hours before you begin the activity, to hasten germination. Place the dried seeds in a container, and add enough water to cover the seeds. (Each student that prepares a "seed rack" will use 4 seeds.) Drain the seeds before you disinfecxt them.

4. To prevent the seeds from getting moldy and not germinating, immerse them in a seed disinfecting solution for 2 minutes. A seed disinfecting solution can be made by mixing ¼ cup (59 mL) of bleach with 1 quart (944 mL) of water. The easiest way to disinfect the seeds is to place them in a strainer, and then dip the strainer and the seeds into the disinfecting solution for two minutes. A straining bag can be also be made by using discarded stockings, or nylon netting. After the

seeds have drained, place them in a clean, labeled container. The seeds are ready to be used immediately.

5. Obtain plastic bags that will easily fit over the folded cardboard. (You can also cover several "seed racks" at a time by inverting an aquarium over them.) You will also need paper towels, scissors to cut the slot out of the cardboard, permanent markers for labeling, several staplers, extra staples, containers for water, and straight pins.

6. Prepare a folded cardboard seed rack, and attach the paper towel. Cut a slot, and attach 4 seeds. Place one half of the cardboard into a container of water. Your students can use this model for reference as they construct their own seed racks.

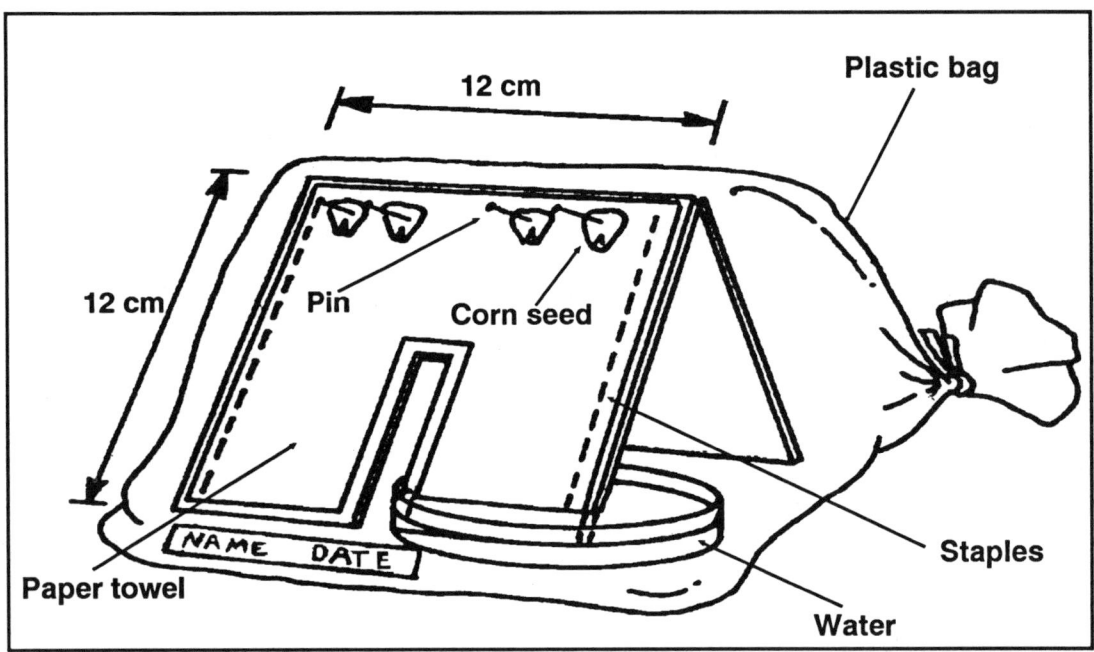

7. Set up the supply table. Neatly arrange the disinfected corn seeds, the cardboard, the paper towels, the staplers, the scissors, the pins, the containers, the markers, and the plastic bags.

8. Provide paper towels and soap for cleaning the work surfaces and hands.

9. Discuss the procedure with your students. Remind them to wash their hands as soon as they have pinned the seeds to the cardboard. Show them where to place their seed racks when they have finished setting them up.

Basic Biology Science Activities 43

TEACHER'S GUIDE
WHY DO PLANTS GROW TOWARDS LIGHT?
Phototropism

GOALS:

1. To define phototropism.

2. To observe the growth of plant stems in response to light.

3. To explain why phototropism is important for plant survival.

SAFETY:

Caution your students that the seed disinfecting solution is a poison, and that they should wash their hands after they handle the seeds. They should wash until their hands no longer feel slippery. Goggles should be worn to protect eyes from the disinfecting solution, and aprons should be worn to protect clothing. You should be sure that your students do not eat the seeds used in this activity. Scissors can be dangerous if used carelessly.

PREPARATION AND HINTS:

1. This activity can be done individually, or in groups of three.

2. Your students can prepare the light–tight boxes that are needed for procedures 11 – 13 by using 2 quart milk cartons. The top of each carton should be cut off, and the cut ends trimmed so that they are even. You can also use large corrugated cardboard boxes, that can cover many containers of growing seeds simultaneously. The boxes can be covered with black paper. A slot should be cut in one of the boxes or in half of the milk cartons.

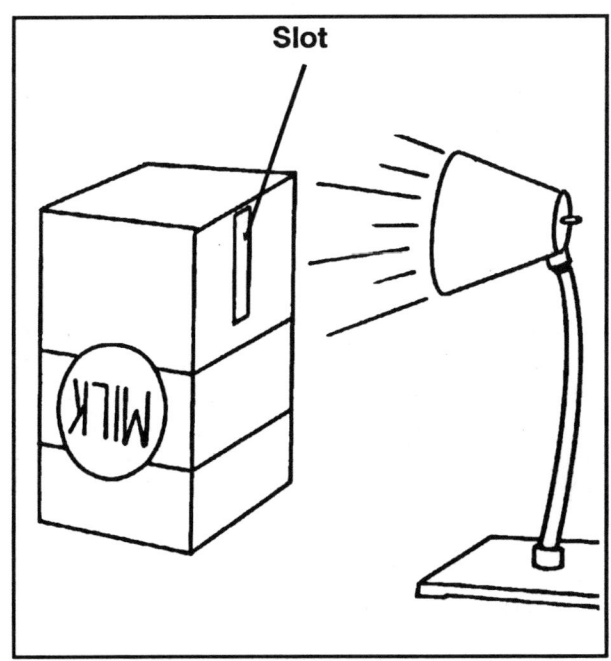

When the plants are covered by the box with the slot, light will enter the box only through the slot. The plant stems will grow towards the slot. This type of plant growth is called phototropism.

3. Corn or bean seeds should be soaked for 24 hours before you begin this activity, to hasten germination. Place the dried seeds in a container, and add enough water to cover the seeds. (Each group of 3 students that does this activity will use 30 seeds.) Drain the seeds before you disinfect them. Quick results will be obtained if you use radish seeds rather than corn or bean seeds. Radish seeds do not have to be soaked for 24 hours, but they should be disinfected.

4. To prevent the seeds from getting moldy and not germinating, immerse them in a seed disinfecting solution for 2 minutes. A seed disinfecting solution can be made by mixing ¼ cup (59 mL) of bleach with 1 quart (944 mL) of water. The easiest way to disinfect the seeds is to place them in a strainer, and then dip the strainer and the seeds into the disinfecting solution for two minutes. A straining bag can also be made by using discarded stockings, or nylon netting. After the seeds have drained, place them in a clean, labeled container. The seeds are ready to be used immediately.

5. Obtain containers (paper cups, or small plant pots) and soil in which to plant the seeds. You will also need markers to label the containers, and newspaper to protect the tables during "planting season".

6. Set up the supply table. Neatly arrange the disinfected seeds, the newspaper, the containers, the soil, and the markers.

7. Provide paper towels and soap for cleaning up the work surfaces and hands. (A pan and brush are useful to clean up soil spills.)

8. Discuss the procedure with your students. Remind them to wash their hands as soon as they have planted their seeds. Show them where to place their containers of seeds until the seeds germinate.

9. Allow time each day for your students to check their seeds to see if they have germinated, or need to be watered.

10. When the seeds have germinated, place the cardboard boxes, or the milk containers, on the supply table. Show your students where to place the seeds that will be in the light (the seeds that will be covered by the box with a slot in it should be placed in the same area), and the seeds that will be kept in the dark. If your light source comes from one side (as light coming in a window) the plants that are not covered should be turned daily so that they do not bend towards the light source.

TEACHER'S GUIDE
WHY DOES A PLANT NEED SUNLIGHT?
Starch Production And Storage In A Leaf

GOALS:

1. To define photosynthesis.

2. To practice the iodine test for starch.

3. To compare the results of the iodine test on leaves from two different plants.

4. To conclude that starch is present only in the leaf taken from the plant that had been kept in the light.

5. To conclude that the plant that has been kept in the dark used up its starch supply.

SAFETY:

Caution your students that the iodine solution is a poison, and that they should wash their hands after they handle the iodine stained leaves. Goggles and apron should be worn. Caution your students that the iodine can temporarily stain their skin, and permanently stain their clothing. The alcohol that is used to remove the green color from the leaf is flammable. It is better to heat the alcohol with a hot plate rather than on an open flame. A double boiler greatly reduces the risk of igniting the alcohol.

PREPARATION AND HINTS:

1. This activity can be done individually, or in groups of two or four. Each pair of students in a group can prepare and test one of the two leaves.

2. Several days before beginning this activity obtain two or more plants. Typical plants are <u>Geranium</u> and <u>Coleus,</u> although other plants that have comparably sized leaves may be used. If plants with very large leaves are used, the leaves may be cut into several pieces before they are decolorized in the alcohol, and tested with iodine. Water your plants well, and keep at least one plant in the dark. This can be inside a closet, or you can cover the plants with several bags or boxes. The

idea is to keep light from striking the leaves of those plants. The other plant(s) should be kept in a brightly lit area. (A lamp may be used to supplement light from the window during winter months).

3. Prepare the dilute iodine solution by mixing 1 teaspoon (5 mL) of medicinal iodine in 1/2 cup (118 mL) of water. Each leaf that is tested will be covered with the iodine solution, so be sure that you have several cups of solution available. Greater quantities of the iodine solution can be prepared by mixing 2 tablespoons (30 ml) of iodine in 3 cups (700 ml) of water. (If you have prepared iodine indicator for a previous activity, you may use it for this activity.) The iodine solution can be placed in small containers and dispensed with eye droppers, or in small squeeze bottles (empty contact lens solution bottles can be used, and capped for storage).

4. Obtain containers (paper cups, small dishes or small beakers) to be used to test the leaves with the iodine solution.

5. Obtain tongs, forceps or spoons that can be used to remove the leaves from the hot water, alcohol and the iodine solution.

6. Obtain bread or potatoes to show that iodine changes from reddish-brown to bluish-black when placed on starch. (If your students have done earlier activities in which they have used iodine as an indicator for starch, your students can omit procedures 2 and 3.) Cut the crusts off the bread and discard them. Cut the remaining bread into small squares, or cut the peeled potatoes into small pieces. Put the bread or potato in a labeled container.

7. Obtain alcohol (rubbing alcohol is effective), containers to make a double boiler, and a hot plate.

8. To see the effect of the iodine indicator solution, the green chlorophyll in the leaves has to be removed. You should do this as your students watch. Fill 2 containers halfway with water. Boil the water, and place the leaves that have been in the light in one container, and the leaves that have been in the dark in the other container. Use tongs or forceps (tweezers) to remove the leaves after 2 – 3 minutes. (Boiling the leaves softens them and makes it easier to remove the green chlorophyll. This step can be eliminated, although the final results will not be as good.) The boiled leaves should be placed in 2 containers that have been filled halfway with alcohol. The containers of alcohol and leaves should be placed in a water bath (a slightly larger container with water that is about 1 inch (2.5 cm) deep), which is then placed on the hot plate. As the water is heated, it will heat the alcohol. If you only have one hot plate, boil and decolorize the "light" leaves first. Use

tongs or forceps to remove the pale green–white leaves from the hot alcohol. The leaves can then be placed in a dish of water to rinse off the alcohol. Your students can then place the leaves in small containers and test them with iodine.

9. Set up the supply table. Place the iodine solution, and container of bread or potato on the table. Arrange the containers and tongs (spoons or forceps) neatly. Newspaper can be provided to cover your students' tables in case of iodine spills. Tape will be needed to attach the iodine stained leaves to the Results and Observations section.

10 Provide paper towels and soap for cleaning work surfaces and hands.

11. Discuss the procedure with your students. Remind them not to touch the iodine with their fingers, and to wash their hands as soon as they have taped the leaves to the paper.

12. Challenge your students at the beginning or end of this activity to design an experiment would produce a pattern on a leaf when it is placed in an iodine solution. (To make a leaf that has a pattern on it, do the following. Four days before you begin this activity, put a plant in the dark. Take it out of the dark 2 days later. The plant has now used up any starch that was stored in its leaves. Cover part of the upper and lower surface of several leaves with aluminum foil that you have cut into a geometric shape, or letter of the alphabet. You can even cover half the leaf with foil, leaving the other half uncovered. Leave the plant in the light for 2 days. Only the surfaces of the leaf that are uncovered will do photosynthesis when the light strikes them. Excess sugar is stored as starch in those areas only. The covered surfaces do not do photosynthesis, so they don't produce sugar or starch. When the foil is removed, boil the leaves in water, then in alcohol, and then test with iodine. At this point you should show the leaves to your students. They will see the original foil cut–outs as dark brown or black areas surrounded by a light tan color.)

48 Fundamentals of Science Activity Series

TEACHER'S GUIDE
CAN YOU GROW A PLANT WITHOUT SOIL?
HYDROPONICS

GOALS:

1. To measure the heights of seedlings grown in different conditions
2. To calculate the average heights of a group of seeds.
3. To compare average heights of seedlings grown with plant minerals, to the average heights of seedlings grown without plant minerals.
4. To conclude that seedlings must be supplied with plant minerals for healthy growth.
5. To review the raw materials needed for photosynthesis.

SAFETY:

Caution your students that they should wash their hands after they handle the seeds and the plant mineral solution. Goggles and aprons should be worn.

PREPARATION AND HINTS:

1. Students can work alone or in pairs. Younger students can do this activity if they omit the measurements and calculations. Written observations and sketches can be substituted for the measurements.

2. Soak the seeds in water for 24 hours prior to the activity to hasten seed germination. (This step may be omitted, but the seeds will take an extra day to germinate.) To soak the seeds, put them in a dish, and add enough water to cover them. You will need 20 seeds for each group of students that does this activity. Drain the water before you disinfect the seeds.

3. To prevent the seeds from getting moldy and not germinating, immerse them in a seed disinfecting solution for 2 minutes. A seed disinfecting solution can be made by mixing ¼ cup (59 mL) of bleach with 1 quart (944 mL) of water. The easiest way to disinfect the seeds is to place them in a strainer, and then dip the strainer and the seeds into the disinfecting solution. A straining bag can also be made using a discarded stocking or nylon netting. After the seeds have drained,

place them in a clean, labeled container. The seeds are ready to be used immediately.

4. You will need two containers for each group that does this activity. You may use cups, saucers, or even plastic margarine containers.

5. Obtain sponges and use scissors to cut them to fit in the bottom of the containers. If sponges are not available, 0.5 inches (1 cm) of clean sand, gravel, or perlite can be placed at the bottom of each container.

6. Prepare the plant mineral solution according to the directions on the package of plant food. (You can use Miracle Gro or a similar product.) Place the solution in several containers labeled "plant minerals". (Squeeze bottles from contact lens wetting solution work well. The mineral solution can be dispensed easily, spills are minimized, and the containers can be capped for easy storage.)

7. Set up the supply table. Neatly arrange the seed growing containers, sponges, and seeds. Nearby place the containers of "plant minerals", markers for labeling, rulers, and newspaper.

8. Provide paper towels and soap for cleaning work surfaces and hands.

9. Discuss the procedure with your students. Remind them to wash their hands as soon as they have placed their seeds on the sponge, again after they have added the plant mineral solution, and a third time when they clean up. Show your students where to place the containers of seeds when they are done.

10. Allow time each day to check the seeds, and to measure them as they grow. The activity can end earlier than the 10 days called for in the procedure if your students can see differences in the appearance of the two groups of seedlings.

TEACHER'S GUIDE
CAN YOU GROW A NEW PLANT WITHOUT USING A SEED?
Vegetative Propagation

GOALS:

1. To define vegetative propagation.
2. To demonstrate the process of vegetative propagation.
3. To observe which appeared first during the process of vegetative propagation, the roots or the stem and leaves.

SAFETY:

If your students are to cut their vegetables by themselves, warn them that they can be injured by knives if they are careless. The toothpicks should be inserted into the vegetable when the vegetable is resting on a hard surface, not when it is held in the hand.

PREPARATION AND HINTS:

1. Students should work individually on this activity.
2. Younger students can do this activity if you push the toothpicks into the vegetable for them. The vegetables can then be placed in a cup of water by a child of any age.
3. Obtain an assortment of vegetables for your students to choose from if they are not going to bring them from home. Among the most easily grown vegetables are carrot, onion, garlic, potato, and sweet potato. If you don't want your students to handle knives, cut the pointed end of the carrots off, and discard them.
4. Obtain containers that hold water and are wide enough for the vegetables that your students will be using.
5. Obtain round or square (not flat) toothpicks.
6. Set up the supply table. Place the vegetables on the table, and provide the name of each vegetable. Place the toothpicks, containers, and markers on the table. If your students are going to cut the end of their carrots, have them use the knives while you watch them.

7. Provide paper towels and soap for cleaning work surfaces and hands.

8. Discuss the procedure with your students. Remind them to wash their hands when they clean up. Show your students where to place their containers of vegetables when they have completed their set-up.

9. Allow time each day to check the progress of the vegetative propagation.

TEACHER'S GUIDE
SHE LOVES ME, SHE LOVES ME NOT. CAN YOU NAME THE PARTS OF A FLOWER?

Parts Of A Flower

GOALS:

1. To identify the parts of a flower.
2. To dissect a flower, and to remove and label the individual parts.
3. To describe the function of each part of a flower.

SAFETY:

Razor blades should be used with care. If your students are too young to handle a razor blade, procedure 8 may be omitted, or you may demonstrate this step for your students.

PREPARATION AND HINTS:

1. Students should work individually or in pairs.
2. Obtain flowers such as gladiolus, daffodil, or tulip. If this activity is done in the spring, many such flowers are available. You can also speak to florists in your area. They will often donate flowers that they would normally discard if you express an interest for your class. Another source of flowers is your local funeral home. If you speak to the director, the floral arrangements that are not kept by the families of the deceased can be saved for you to bring to class. (Before you bring the flowers to class break the arrangement up and bring the flowers that you will dissect in a large container filled with water.)
3. Obtain hand lenses or magnifying glasses. These will be helpful when your students are to observe the pollen on the anther, and the ovules in the ovary.
4. Obtain tape to secure the parts of the flower to the appropriate sections in the Results and Observations.
5. Set up the supply table. Place the flowers on the table in a bowl or vase of water. (You may want to break gladiolus flowers off the stalk, and float the individual flowers in a bowl of water.) Place the tape, rulers, and magnifying glasses on the table. Keep the razor blades in

a secure area if your students are to cut the ovaries themselves, and have them use the razor blades while you watch them.

6. Provide paper towels and soap for cleaning work surfaces and hands.

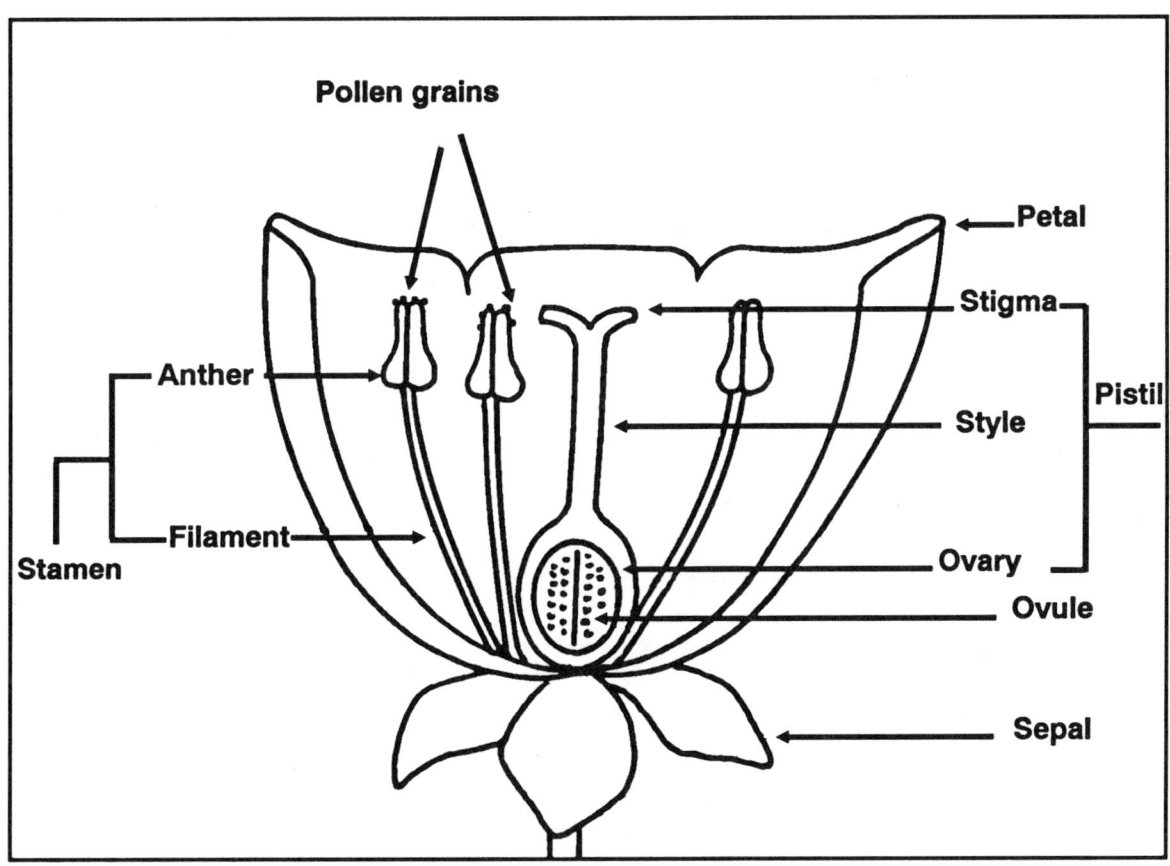

TEACHER'S GUIDE
CAN YOU CATCH A FALLING RULER?
Measuring Reaction Times

GOALS:

1. To determine how long it takes for you to catch a falling meter stick.
2. To read a chart to determine your reaction time.
3. To calculate your average reaction time.
4. To prepare a chart that organizes the average reaction times.

SAFETY:

Check the meter sticks for rough edges that could cause splinters. Have a piece of sandpaper available to smooth such edges.

PREPARATION AND HINTS:

1. Students should work in pairs.
2. Obtain 1 meter stick for each pair of students. (If you do not have meter sticks, you can use yard sticks. If you divide each number on the data chart by 2.54 you can convert the lengths which are given in centimeters into lengths in inches. The values are approximately 2 inches, 4 inches, 6 inches, through 40 inches.)
3. Train your students that meter sticks are measuring devices. Their accuracy depends upon the condition they are in. Meter sticks should not be banged on the table, or swung like bats.
4. Set up the supply table by laying out the meter sticks.
5. Discuss the procedure with your students. Be sure they understand that the person standing holds the meter stick at the 50 cm mark (20 inches), and raises or lowers the meter stick so that the 0 centimeter mark (lower edge of the meter stick) is at the fingertips of the partner who is sitting.

TEACHER'S GUIDE
LISTEN TO YOUR HEARTBEAT

Making A Stethoscope To Measure The Rate Of Your Heart

GOALS:

1. To construct a simple stethoscope and use it to listen to your heartbeat.
2. To measure your heart rate (pulse).
3. To compare your heart rate with your partner's heart rate.
4. To calculate the number of times your heart beats in 1 hour.

SAFETY:

Caution your students that they should hold the end of their stethoscope **near** their ear, and not insert it **into** their ear.

PREPARATION AND HINTS:

1. This activity should be done in pairs.
2. Obtain the supplies to construct the stethoscopes. Each pair of students will need one stethoscope. Surgical tubing can be used for the hose of the stethoscopes, and is available in pharmacies and stores that sell home health care products and medical supplies. If your students might place the end of the hose into their ears, obtain two funnels for each stethoscope. (One funnel is placed over the heart, and the other funnel over the ear.)
3. Cut the hoses for the stethoscopes into approximately 3 foot (1 meter) lengths.
4. Prepare a sample stethoscope. If it is difficult to slide the hose over the end of the funnel, glycerine can be used as a lubricant. Tape the tubing to the funnel(s) if they slide off easily. If your students are very young, you might have to make the stethoscopes for them.
5. Set up the supply table by arranging the rubber hoses, funnels, and tape. If you use glycerine as a lubricant, provide a small amount for your students.

6. Discuss the procedure with your students. If they are to make stethoscopes with a funnel on each end of the hose, be sure to tell them to get two funnels from the supply table. After they have constructed their stethoscopes, tell your students to be quiet. It is very hard to hear the sound of a heartbeat if there is a lot of background noise and talking.

TEACHER'S GUIDE
RACE AGAINST YOUR HEART

Can You Work As Well As Your Heart Can?

GOALS:

1. To transfer water from one container to another as rapidly as your heart pumps blood through your body.

2. To observe that you are not able to maintain a rapid, steady transfer rate for long periods of time.

3. To conclude that the pumping ability of your heart is amazing.

SAFETY:

The transfer of water from one container to another is likely to result in wet tables and floors. Caution your students to spill as little water as possible, and to mop up any spills as soon as they finish each 1 minute trial. Wet floors can be slippery, and lead to accidents.

PREPARATION AND HINTS:

1. This activity should be done in pairs.

2. This is a wet activity. You will need many newspapers and paper towels (and a mop) if you are planning to do this indoors. You will probably want your students to wear aprons. This activity can be done outside, and is very effective during the last week of the school year.

3. Obtain 2 large bowls or pots for each pair of students. The size of the bowls depends upon the size of the transfer device that you choose to have your students use. If you use measuring cups or ladles that hold about ¼ cup (59 mL), you will need large bowls that can hold at least 1 gallon (3.8 liters) of water. (Your students will make about 60 transfers in 1 minute.) If you use tablespoons, the bowls can be smaller.

4. Obtain the transfer devices. You will need one measuring cup (¼ cup), ladle, or tablespoon for each pair of students.

5. If your students are going to do procedure 8, you will need several large containers to measure the volume of water transferred. Two quart pitchers are satisfactory. (Procedure 8 can be omitted.)

6. Prepare the supply table by neatly arranging the bowls, transfer devices, measuring containers (for procedure 8), newspapers, and paper towels.

7. Provide a large container for the disposal of all the wet newspaper and paper towels.

8. Discuss the procedure with your students.

Basic Biology Science Activities 59

TEACHER'S GUIDE
DISAPPEARING ACT
Finding Your Blind Spot

GOALS:

1. To locate your blind spot.
2. To identify the iris and the pupil.

SAFETY:

Caution your students not to hold the ruler directly in front of their partner's eye when measuring distances. It is much safer to hold the ruler to the side of the face.

PREPARATION AND HINTS:

1. This activity should be done in pairs.

2. Although this activity is simple, some younger students have a difficult time staring directly at one letter (on the blind spot test card) while noticing that the other letter on the side of the card has disappeared from their sight. As soon as they shift their eyes to look for the missing letter, the letter reappears because the image no longer falls on the blind spot. For this activity to work, your students must be able to stare directly at one letter, while still being aware of the other letter.

3. You can prepare the blind spot test cards yourself, or have your students prepare them. If your students are particularly fond of magic, you can tell them that they can "magically" make one of the letters disappear as they move the test card away from their face. If they move their eyes, or the card, the letter will reappear "magically". It is not necessary for the blind spot test cards to have the letters **R** and **L** on them, you can substitute other letters, symbols, or small pictures. Be sure that your students understand that they stare at the symbol that is on the **opposite** side of the card from their open eye. (For example, if their right eye is uncovered, they stare at the symbol on the left side of the blind spot test card.)

4. Each pair of students could make their own original blind spot test card, as long as the size of the symbols are comparable to those on

the sample test card. (If the symbols are too large, only part of the image will fit on the blind spot, and so the entire image will never totally disappear during testing.)

5. Set up the supply table by arranging blank paper or index cards and markers (or the completed blind spot test cards), and the rulers.

6. Discuss the procedure with your students. If they can't find their blind spot, watch their eyes as they move the card away from their face. Chances are, they are moving their eyes, or not staring at the correct letter. (The open eye stares at the letter or symbol on the opposite side of the test card.)

Basic Biology Science Activities 61

TEACHER'S GUIDE
CATCH A GLIMPSE

Measuring Peripheral Vision

GOALS:

1. To measure the peripheral vision on each side of your body.

2. To calculate the average peripheral vision on each side of your body.

3. To compare the average values that you obtained for peripheral vision with those of your classmates.

SAFETY:

Caution your students not to move the toy cars too close to their partner's face during the testing.

PREPARATION AND HINTS:

1. This activity should be done in pairs.

2. Although this activity is very simple, good results are obtained only if the student being tested stares straight ahead at the cylinder. The student being tested cannot shift their eyes to the sides during the test.

3. Obtain 2 meter sticks (or yard sticks) for each pair of students.

4. Obtain a toy car or small object for each pair of students. (You could have students bring in their own objects, as long as they are all about the same size.)

5. Obtain modeling clay. Each pair of students will need a piece that is the size of a walnut to roll into a cylinder. The cylinder of clay provides a fixed object for the person being tested to stare at. Other objects that will remain in position can be substituted for the clay.

6. Set up the supply table. Place the meter sticks and tape (to hold the meter sticks in position) in one area, and the clay and toy cars in another area.

7. Discuss the procedure with your students. If they are having a hard time finding where their peripheral vision begins, remind them that they must stare straight ahead at the clay cylinder, and not move their eyes to the side.

62 Fundamentals of Science Activity Series

TEACHER'S GUIDE
THE EYES HAVE IT
Depth Perception

GOALS:

1. To describe binocular vision.

2. To demonstrate how difficult it is to judge distances when one eye is closed.

3. To compare the ability to judge the distance between two objects when both eyes are open, and when one eye is closed.

SAFETY:

If your students are going to make their own depth perception viewing cards, remind them that scissors should be used carefully.

PREPARATION AND HINTS:

1. This activity should be done in pairs.

2. If possible, obtain several stereoscopic viewers to demonstrate depth perception. They are available in toy stores, or your students may have them at home. Each scene on the viewing card is composed of two separate photographs. Your students should hold the viewing cards up to the light and look at the individual photographs closely. Pairs of photographs will appear to be identical. In actuality, they are pictures of the same scene taken from slightly different angles. When the viewing card is placed in the viewer, each eye sees a slightly different view of the scene, which makes the scene appear to be in 3D. Some images jump to the foreground, while others drop into the background. Your students will be able to tell which objects are in the front, and which are in the back. If they look at the viewer with one eye closed, or covered, they will find that the picture looks flat, or 2D, and that it doesn't look as if some objects are in front of other objects.

3. You can do a simplifed version of this activity with younger students. Have one person put their chin on the table, and look straight ahead. A second person will hold two similar objects on the table about 3 feet away from their chin. One object should be placed slightly in front of

the other object. The first person has to say which object is closer to their chin. This is easy when both eye are open. A correct response can be recorded on a data chart as a (+), while an incorrect response can be recorded as a (−). This entire process should be repeated several times with both eyes open, and then several times with the left eye covered, and then with the right eye covered. There should be many more correct responses when both eyes are open. The students should switch positions and repeat the activity.

4. Obtain 3" x 5" index cards and scissors. You can cut the viewing slots for your students if you want to save class time, or if you don't want your students to handle scissors.

5. Obtain modeling clay or spring clothespins, and pencils or dowels. You will need two for each pair of students.

6. Obtain one meter stick (yard stick) for each pair of students.

7. Obtain blank paper to act as the background for the depth perception test.

8. Set up the supply table by neatly arranging the materials that you have gathered.

9. As you discuss the procedure with your students, set up one table and demonstrate the procedure with a pair of students. You could set up the table before you begin the activity, or as you talk to your students.

TEACHER'S GUIDE
PUPILS THAT DON'T GO TO SCHOOL
Reflex Reactions Of The Pupils Of Your Eyes To Light

GOALS:

1. To identify the iris and pupil of the eye.
2. To observe that the pupil contracts when a bright light is shined on it.
3. To sketch the pupil and iris in both low light and bright light conditions.
4. To describe what happens to the size of the pupil when it is exposed to a bright light.
5. To describe the function of the pupil reflex.

SAFETY:

Caution your students not to bring the flashlight too close to their eyes, or too close to their partner's eyes.

PREPARATION AND HINTS:

1. This activity can be done individually, or in pairs.
2. Each student, or pair of students that is doing this activity will need a flashlight. (If mirrors are available, students can look at their own pupils contract.)
3. Place the flashlights and mirrors on the supply table.
4. As you discuss the procedure with your students, darken the room so that their pupils begin to dilate. You can give an example of a reflex by hitting a table or desk top with the flat side of a meter stick (yard stick). The unexpected loud noise will make many or all students flinch.

Basic Biology Science Activities 65

TEACHER'S GUIDE
WHAT'S THE BEST WAY TO EAT A LOLLIPOP?

Find The Sweet Taste Receptors On Your Tongue

GOALS:

1. To identify the area of the tongue that contains the most taste receptors for each of the four tastes.

2. To list the four tastes that the tongue can detect.

3. To map the individual areas of taste on the tongue.

SAFETY:

Caution your students that they should use each cotton swab only once. Repeated use will contaminate the test solutions with bacteria from their mouths, and with other test solutions that may be on the cotton swab.

PREPARATION AND HINTS:

1. This activity can be done in pairs, or in groups of up to 4.

2. You will need approximately ¼ cup of each of the 4 test solutions for each pair or group of students who does this activity. To prepare the sweet solution dissolve 2 teaspoons of sugar in each cup of water. The sour solution is prepared by mixing equal amounts of vinegar and water. The salty solution is prepared by mixing 4 teaspoons of salt in each cup of water. The bitter solution can be strong, black coffee, or you can dissolve 2 aspirins in each cup of water.

3. The solutions should be put in small, clean, labeled containers that can be covered (and refrigerated if you are preparing these in advance) until needed. Baby food jars make ideal containers because they are small, and have covers. Your students can probably bring in a large supply of empty jars. You will need one set of 4 jars for each group of students. If your students are older, or need practice in measuring, you can have them dispense the solutions from a large container on the supply table into small containers, as described in procedure 1.

4. Your students will need graduated cylinders or tablespoons to measure the solutions in procedure 1. (This step can be omitted to save

time. Prepare the jars for each group before class, and have students begin with procedure 2.)

5. Obtain colored pencils or crayons, cotton swabs, and tissues.

6. Place the solutions to be tasted on the supply table. Arrange the jars so that similar solutions are grouped together. Neatly arrange the crayons, tissues, and cotton swabs.

7. Discuss the procedure with your students. Stress that they **must** use a fresh cotton swab every time they dip the swab into the test solution. (This is to prevent contamination of the test solution by other test solutions, or by bacteria from the mouth.) To be sure of accurate diagrams, your students should carefully dry their tongue with tissues before each taste test.

TEACHER'S GUIDE
HOW SENSITIVE ARE YOU?

Measuring The Accuracy Of The Sense Receptors In Different Parts Of Your Body

GOALS:

1. To test different areas of the skin to identify the area that has the most touch receptors.

2. To compile a chart that summarizes the class data on the part of the body that is most sensitive to touch.

SAFETY:

Tell your students that they should use the toothpicks to touch the skin gently. They should not press the toothpicks down into the skin. The toothpicks are being used to measure the pressure receptors, and not the pain receptors.

PREPARATION AND HINTS:

1. This activity should be done in pairs.

2. You will need 1 ruler, 2 round toothpicks, and two small rubber bands for each pair of students.

3. To save time in class, you can prepare the testing devices for your students. If they are to construct the testing devices themselves, have a sample prepared, and demonstrate the technique as you discuss the procedures in class. To make the testing device work, the toothpicks must both rest on the skin at the same time. One toothpick can't touch the skin before the other. (The level of the tooth picks can be adjusted by touching them to a flat surface.)

4. If you need to save time, have your students test just a few areas of the body. (The forearm and palm of hand would be the best sites to choose if you are testing only two areas.)

5. Place the rulers, toothpicks and rubber bands (or completed testing devices) on the supply table.

6. Discuss the procedure with your students. Stress that they are mapping the pressure receptors, and not the pain receptors. The tooth

picks should rest **lightly** on the skin, and not leave marks on the skin. The person being tested should not look at the testing device. They should have their eyes closed, or be blindfolded during testing.

TEACHER'S GUIDE
ARE YOU FULL OF HOT AIR?
How To Measure Your Lung Capacity

GOALS:

1. To define vital capacity.
2. To determine the vital capacity of your lungs.
3. To compare your vital capacity with your partner's.

SAFETY:

This is a very wet activity. If water spills on the floor, mop it up to prevent slips.

PREPARATION AND HINTS:

1. This activity should be done in pairs.
2. This can be a very wet activity. Your students should wear aprons, and you should have newspaper, paper towels, and a mop handy.
3. Obtain a large plastic bottle (3 liter soda bottle, or a 1 gallon plastic container) for each pair of students.
4. Calibrate the bottles by pouring 1 cup (236 mL) of water into one of the bottles. Mark the level of the water on the outside of the container. Continue to add water a cup at a time, marking the water level after each addition, until the container is full. If the other bottles are identical, you can then use that bottle as a guide to mark bottles for the rest of the students. (You will have to calibrate each different type of bottle.)
5. Obtain a deep pan for each pair of students that can hold at least a gallon of water. Plastic wash basins or disposable roasting tins can be used for this activity.
6. Obtain a funnel and a cup for each pair of students.
7. Each student will need a flexible length of hose that is about 2 feet long. Drinking straws that have a flexible joint can also be used.

8. Arrange the calibrated containers, the large pans, funnels, cups, and the hoses or straws on the supply table. In addition, have paper towels, newspapers, and mops available to clean up any spills.

9. As you discuss the procedures with your students, you can mention that they are using the water displacement method to determine their vital capacity. As they exhale into the hose (or straw), bubbles of their breath will gather in the large calibrated container. Some of the water in the container will be forced out (displaced), and be replaced by their exhaled air. The volume of the water that is displaced is equal to the volume of their exhaled breath, their vital capacity.

Basic Biology Science Activities 71

TEACHER'S GUIDE
INHALE, EXHALE, INHALE, EXHALE
Using Limewater To Test For Carbon Dioxide In Your Breath

GOALS:

1. To use limewater to test for the presence of carbon dioxide.

2. To observe that exhaled air tests positively for carbon dioxide.

3. To conclude that exhaled air contains more carbon dioxide than the air that is inhaled.

SAFETY:

Your students should be careful not to aspirate the limewater as they perform the limewater test. They should wear goggles to protect their eyes, and aprons to protect their clothing from splashes of limewater.

PREPARATION AND HINTS:

1. This activity can be done in groups of up to four.

2. Prepare and test the limewater several days in advance. (You will need about 1 cup of limewater for each of your students who performs the limewater test.) Mix warm water with your source of lime, and allow the lime to settle overnight. The next day, carefully pour off, and save, the clear liquid. This clear liquid is your limewater. Discard the lime. Test a small amount of the limewater to be sure that it turns cloudy (positive test) when you exhale into it through a straw. (Blow gently so that you don't splatter limewater.) If the limewater does not get cloudy after a few minutes of blowing bubbles into it, discard the solution, and find a new source of lime. If your limewater tested positively, the remainder can be stored in a sealed and labeled container until needed.

3. Obtain clear cups, beakers, or test tubes. You need at least 1 container for each student, although the activity will go faster if the students have up to 4 containers, and do not have to rinse a single container at the end of each test.

4. Each student will need a straw, and a spoon for measuring the soda.

5. You will need seltzer (or other light colored soda), and several cooking basters, rubber syringes, or eye droppers.

6. Place the limewater, straws, spoons, containers, and basters on the supply table.

7. Discuss the procedures with your students. The limewater test is an irreversible reaction. (Once the limewater turns cloudy, it will not get clear again.) Make sure that your students don't use cloudy limewater for any of their tests.

TEACHER'S GUIDE
CAN YOU DO IT?

Experiments In Balance

GOALS:

1. To find the center of gravity.

2. To observe that balance can be maintained only if the center of gravity is directly over the center of support.

SAFETY:

Your students should be careful as they try to balance on one foot that they don't fall and hurt themselves.

PREPARATION AND HINTS:

1. This activity can be done individually, or in groups of up to four. (It is more fun when done individually, though.)

2. Obtain objects to locate the center of gravity. Plastic bags filled with 1 cup of sand or gravel will work well, or you can use any other nonbreakable object that can hang from a string. You will need one object for each student, or group, that does this activity.

3. Obtain rope or string cut into pieces that will tie around your student's chests on one end, and to the object at the other end. The rope should be long enough to suspend the object slightly above knee level. (Your rope (or string) will probably be about 5 feet (1.5 meters) long.)

4. Tie the pieces of rope or string securely to each object, and place them on the supply table.

5. Discuss the procedure for Part One with your students. Instruct them to work in an area that is cleared of chairs and tables. Your students might be tempted to hold onto the chairs to balance in odd positions. (They will find that if they **don't** hold on to anything, they can only balance if the object hangs over the one foot (or between the two feet) that they are standing on.) Another reason to keep your students away from furniture is that if they do lose their balance and fall, they are less likely to injure themselves.

6. After your students have experienced the importance of keeping the center of gravity over the center of support to maintain balance, challenge them with Procedure Part Two. They must stand with their back and heels against a wall and pick up a piece of paper, without falling forward. You can use one of your students to demonstrate this, but be aware that they will **not** be able to maintain their balance, and that they **will** fall forward. Your students can try this activity without the object tied around their chests if they think it is causing them to fall. They will still fall!

7. Further reinforcement of the center of gravity with respect to balance occurs in Procedure Part Three. It sounds easy to stand on tiptoes, but with your toes, body and face against the edge of an open door, it can't be done. Use one student to demonstrate this, then let everyone try it.

Basic Biology Science Activities 75

TEACHER'S GUIDE
CAN YOU BEND A BONE OR BOUNCE AN EGG?
Using Vinegar To Remove Calcium Compounds

GOALS:

1. To predict what will happen to an egg or a bone that is placed in vinegar.

2. To observe the production of a gas (carbon dioxide) when a bone or an egg shell is placed in vinegar.

3. To demonstrate that bones and egg shells are composed primarily of calcium compounds (calcium phosphate, and calcium carbonate).

4. To handle a bone and an egg that have had the calcium compounds removed.

5. To conclude that most of the strength of an egg or a bone comes from calcium carbonate and calcium phosphate.

SAFETY:

Your students should wear goggles and aprons to protect themselves from the vinegar. They should wash their hands after they handle the eggs and bones that have been soaked in vinegar.

PREPARATION AND HINTS:

1. This activity can be done individually, or in groups of up to four.

2. Obtain raw eggs, and chicken leg or wing bones for each of your students. (After the chicken has been cooked, boil the bones to remove any small pieces of meat that remain on the bones.)

3. Obtain at least 1 quart (1 liter) of vinegar for each group doing this activity.

4. Obtain 4 containers for each group of students. The two containers that will hold the eggs should be large enough so that your students are able to reach into the containers and cup the egg in their hand to remove it from the container. The containers for the bones should be large enough so that the bone can be completely covered with vinegar.

5. Set up the supply table for the "rubber egg" procedure. Place the vinegar, containers, raw eggs, and markers for labeling on the table.

6. Discuss the procedure with your students. It will take a minimum of 24 hours for the calcium carbonate to be removed from the egg, and this is if the egg is rubbed gently with a tissue periodically to remove the softened material. You can set the activity up on a Friday, and wait until Monday to remove the egg from the vinegar.

7. When it is time for your students to remove the eggs from the vinegar, have them spread paper towels or newspaper on their tables, and wear their aprons and goggles. (Several of your students will probably squeeze the eggs too hard, causing the eggs to break. Cleanup will be easier if you plan ahead!) Your students will need tissues to wipe off the soft white material from the membrane of the egg. The eggs can be **very gently** bounced, but it should be obvious that the original shell is what protected the contents of the egg. The eggs (whole and broken) should be placed in a separate bag for neat disposal.

8. Set up the supply table for the bone bending activity. Include two containers, bones, vinegar, and markers for labeling.

9. Rather than discuss the procedure for making a bone that bends, challenge your students to apply the information they learned about vinegar and calcium in the "rubber egg" activity to make a bone that bends.

10. When your students check their bones the next day, they will find that the bones don't bend. Reassure them, saying that bones are much thicker than egg shells, and that they should have confidence in their experimental designs.

11. Have your students change the vinegar every 2 – 3 days. It will take about 10 days to produce a bone that bends.

Basic Biology Science Activities 77

TEACHER'S GUIDE
CAN YOU MAKE YOUR OWN YOGURT?
Making Yogurt From Milk

GOALS:

1. To observe that yogurt is a milk product.

2. To conclude that the bacteria in a small amount of yogurt are able to change a relatively large quantity of milk into yogurt.

SAFETY:

Students should wear aprons to protect their clothes. If they are to taste the yogurt that they make, all cups and spoons must be new, and clean. Stress that your students wash their hands before making and tasting the yogurt.

PREPARATION AND HINTS:

1. This activity can be done individually, or in pairs.

2. Prepare a warm area to serve as an incubator for the yogurt bacteria. You can cover a heating pad that is set on low with a box, use an oven that has a gas pilot light or has the oven light left on, or put the containers on the warm top of an appliance. As a last resort, you can pour the warm milk into large mouth insulated containers. If the milk is kept at room temperature (rather than in a warm environment), the milk will spoil, rather than become yogurt.

3. Obtain at least ¼ cup of milk for each pair of students, and a container of yogurt (plain is the best flavor to use). This yogurt is used as the source of bacteria and **must** have an ingredients label that says "active cultures", or the names of the bacteria.

4. Each pair of students will need a small, new, clean container and spoon. Paper cups and plastic spoons are best. If your students are to taste the yogurt, it is **essential** that these items be clean and new. Stress that it is only because they are using clean, new containers and spoons, that they can taste their homemade yogurt.

5. Obtain a cooking pot and cooking thermometer (optional), as well as plastic wrap, paper towels, and jam (optional).

6. Carefully wipe the supply table, and place the plastic wrap and paper towels on the table. Place the cups and spoons on the table. (If possible, leave the cups and spoons in their original packaging to show that they are brand new.) Provide jam to use as flavoring (optional) and markers to label the cups.

7. Discuss the activity as you heat the milk. Scald the milk (heat it to 180°F, or until little bubbles form around the edge of the pan), and allow it to cool to 105 – 110°F. (While the milk is cooling, have your students get a cup and write their name on it. If jam is available, they can place a teaspoon of jam in the cup at this time.) Have a volunteer add 2 tablespoons of yogurt to each quart of warm milk, and have another volunteer stir the milk carefully. Carefully fill each cup halfway with the warm milk and yogurt mixture. Tell your students where to bring their cups after they have covered them with plastic wrap.

8. Allow time for your students to check the progress of the yogurt making. If this is the first activity of the day, the yogurt might be ready for a taste test at the end of the day. (The yogurt should be refrigerated after 8 – 10 hours of incubation to prevent spoiling.)

TEACHER'S GUIDE
THERE'S A FUNGUS AMONG US
Growing Mold

GOALS:

1. To demonstrate that dust contains spores that can grow into different types of molds.

2. To observe that mold spores require moist conditions to grow.

SAFETY:

Students should not open their plastic bags to closely examine the mold that they have grown. Some students may be allergic to the mold spores. In addition, it is possible for spores of undesirable fungi to grow on the bread. To prevent contamination, instruct your students to observe the mold through the clear plastic bag without opening the bag.

PREPARATION AND HINTS:

1. This activity can be done individually, or in pairs.

2. Prepare an area that can be used to grow the mold. The area should be dark, and at room temperature. (A drawer or cupboard that is opened infrequently is ideal.) The area should be large enough that the bags containing bread can be stored in a single layer.

3. Obtain two plastic, ziplock sandwich bags for each pair of students. Each pair of students will also need a slice of bread, and two paper towels.

4. Half of each slice of the bread will be dampened. The easiest way to assure that the bread is damp, and not soggy, is to provide several spray bottles of water. (You can use empty spray bottles that have been carefully washed.)

5. Arrange the materials on the supply table. Provide markers to be used to label the bags.

6. As you discuss the procedure with your students, show them how to dampen the bread. If you use spray bottles, several squirts will be sufficient to dampen the bread. If you do not have spray bottles, your

students should get their hands wet, and then gently rub their wet hands on the bread.

7. Tell your students where they should wipe their bread to collect dust, and where they should store the dusty bread to allow the spores to grow. Remind them to write their names on each bag.

8. Five to seven days later, have your students examine their bread slices. (You should check the progress of the mold daily to decide the day your students should observe the bread.) Stress that they are **not** to open the bags to get a better look at the molds that grew. The unopened bags of bread and mold should be discarded.

Basic Biology Science Activities 81

TEACHER'S GUIDE
YEAST POWER
Using Yeast To Make Bread Rise

GOALS:

1. To demonstrate that yeast produces bubbles of gas (carbon dioxide) in bread dough.
2. To observe that the bread dough rises because of the carbon dioxide gas that is produced by the yeast.
3. To determine how much a sample of dough rises during two, half hour intervals.
4. To determine how much a sample of dough rises one hour after it has been punched down.
5. To explain how yeast makes bread dough rise.

SAFETY:

Your students should not taste the dough.

PREPARATION AND HINTS:

1. This activity should be done in pairs.
2. Prepare a warm area that can be used to allow the bread to rise. You may use a heating pad (set on low) covered by a box, or use an oven that has the light turned on, or place the dough in an insulated cooler that has a container of hot water inside.
3. Obtain a mixing bowl (large plastic margarine or whipped topping containers can be used), spoon, and two **clear** plastic cups for each pair of students. Each pair of students will also need two damp paper towels.
4. Obtain yeast, flour and sugar. (One tablespoon of sugar and one package of dry yeast (½ cake of compressed yeast), are needed for each 3 cups of flour. The amount of yeast, flour and sugar that you need depends upon the number of student pairs that you have.)
5. Obtain several measuring cups and measuring teaspoons.

6. Warm ½ cup of water for each pair of students. You can keep the water warm in an insulated container.

7. Arrange the materials on the supply table. Provide markers to be used to label the cups, and newspapers to spread on the tables to reduce the mess. Your students will need rulers and a second color marker to use after the dough rises.

8. Discuss the procedure with your students. Show them how to mark the level of the dough on the outside of their cups with a marker (or a piece of tape). Tell your students where to place their covered cups of dough to rise.

9. Allow time to measure the height of the dough at the specified time intervals.

10. Have your students discard the dough and cups when they have completed the activity.

TEACHER'S GUIDE
MAKING FRIENDS WITH AN EARTHWORM
Measuring The Pulse Of An Earthworm

GOALS:

1. To identify the anterior, posterior, dorsal, and ventral sides of an earthworm.

2. To measure the pulse of an earthworm.

3. To appreciate that the worm can be a fascinating living creature to study and that the worm (like all living creatures) must be treated humanely.

SAFETY:

For the safety of the worm, provide a container of moist soil and leaves if you are not going to use your worms immediately. The worms must be kept moist, and must always be held with moist fingers and hands. The worms should not be squeezed or dropped. Your students should wash their hands when they have finished this activity.

PREPARATION AND HINTS:

1. This activity should be done in pairs.

2. No one should be forced to touch the worms. It is best to pair students up so that one of the students in each pair doesn't mind (or likes) touching the worm. Your students will be able to tell if you are uncomfortable with the worms, and may adopt your attitude. If working with the worms is a problem for you, get someone to help you. Sources of help can range from a cooperative student in your class to a former student still in your school, or to a high school student who is willing to help. Most students don't mind the worms when they understand that the worms feel slippery because their skin must be moist to allow them to breathe, and that they feel cool to the touch because the worms are the same temperature as the cool soil.

3. Obtain a supply of earthworms (also called <u>Lumbricus terrestris</u>, night crawlers, and fish worms, or angleworms). These can be obtained from rich, warm, moist garden soil, in bait stores, or from biological supply houses.

4. To maintain your earthworms until you use them in class, place them in a culture jar that is made from a gallon (4 liter) jar which is two-thirds full of a mixture of 1 part garden soil to 2 parts of dried, brown leaves. The culture mixture should be damp. (If it is too wet, the earthworms will crawl to the surface, and could actually drown.)

5. When you complete activity, the earthworms should be released into a garden.

6. To shorten this activity for younger students, you can omit procedure 7, determining the pulse of the earthworm. The students will still feel and hold an earthworm, and identify the dorsal, ventral, anterior, and posterior surfaces, as well as observe the clitellum, and feel the setae.

7. Obtain a dish, tray, or pan for each pair of students. Disposable foil baking pans work well, as do clean trays from meat and frozen food packages.

8. The supply table should be set up with the earthworms, the trays, and paper towels. At least 1 watch or clock that can be used to time one minute intervals is needed. If a large clock is visible, each team can work at their own pace.

9. Discuss the procedure with your students. Stress that the earthworm is a living creature. It must be handled gently, and always be kept moist. An earthworm breathes through its skin. The membrane of the skin must be moist to permit the exchange of oxygen and carbon dioxide. (The moist skin is why worms feel "slimy" to some people.) The dorsal side (upper surface) of the worm is usually smoother and rounder that the ventral side (lower surface) of the worm. The worm should be placed in the pan that the students have prepared with the dorsal side up. (The worm will usually twist around on its own so that the dorsal side is up.)

10. Some worms are very active, making it difficult to see the pulse in the dorsal blood vessel. Tell your students to cover the worm's head with a moist piece of paper towel. If this doesn't quiet the worm down, they should exchange their worm for a calmer specimen.

11. The worms should be returned to the culture jar when your students have finished. Assure your students that the worms will be given a good home in someone's garden.

TEACHER'S GUIDE
DO BACTERIA LIVE IN THE SOIL?

Using Methylene Blue To Detect Soil Bacteria

GOALS:

1. To observe that the color of methylene blue solution changes when combined with soil that contains bacteria.
2. To identify the control in this activity.
3. To infer that bacteria are present in soil even though they cannot be seen.
4. To describe why soil bacteria are important in the recycling of dead organic matter.

SAFETY:

Goggles should be worn to protect your eyes, and those of your students. Aprons will protect clothing from spills and stains. Your students should immediately notify you in case of a spill or breakage. Your students should wash their hands carefully at the end of this activity.

PREPARATION AND HINTS:

1. This activity can be done individually, or in groups of up to four.
2. Place methylene blue solution in several small containers. Label each container with the words "Methylene Blue Solution". The solution can be dispensed with eye droppers. You can also use small plastic squeeze bottles. (Small contact lens bottles work well. These containers should not be used for any other purpose after they are used for methylene blue. They can be capped and stored for other activities that use methylene blue.)
3. Place the oil in several containers. Plastic containers that have small openings (like contact lens wetting solution containers) reduce the number of spills, are easy to use, and can be capped and stored.
4. Obtain soil from two different locations. Place the soil in containers that are labeled to show where the soil came from.

5. Obtain 3 containers for each group of students. You may use clear plastic cups, baby food jars, beakers, or test tubes.

6. Obtain teaspoons and sugar. You will need 3 teaspoons (1 tablespoon) of sugar for each group of students.

7. Prepare the supply table by first covering it with newspaper. (It will make the clean up of spilled soil easier.) Neatly arrange the soil samples, containers, methylene blue, sugar, and oil on the table. Place several teaspoons into each container of soil, and into the sugar. Provide markers or tape for labeling the containers, and paper towels for cleaning up.

8. Discuss the procedure with your students. Tell them where to leave their containers when they have finished setting them up.

9. Allow time during the next day to observe and record the results of the activity.

TEACHER'S GUIDE
MAKING TRACKS

Using Plaster Of Paris To Preserve Animal Tracks

GOAL:

1. To preserve animal tracks.

SAFETY:

An apron should be worn to protect clothing. Unused Plaster of Paris should be allowed to harden, then discarded in the garbage. Wet Plaster of Paris should **not** be poured down the sink. Your students should wash their hands carefully at the end of this activity.

PREPARATION AND HINTS:

1. This activity can be done individually, or in pairs.

2. Obtain Plaster of Paris powder (calcium sulfate) from a hardware or craft store. Depending upon the size of the tracks you plan to make casts of, you will need 1 - 3 cups of Plaster of Paris powder for each team of students.

3. Collect empty cardboard or plastic milk cartons or soda bottles. Slice several cartons (like a loaf of bread) making 1 - 3 inch (3 - 7 cm) slices to serve as frames to go around the tracks. The other containers (one for every team), will be used as mixing containers for the Plaster of Paris.

4. Gather sturdy sticks to be used for stirring the Plaster of Paris powder and water. (Tablespoons can be used, but will have to be wiped off immediately after stirring the Plaster of Paris.) Place 1 stick or spoon in each Plaster of Paris mixing container.

5. Several containers of talcum powder or baby powder will be needed. Each team will dust the track with powder before the Plaster of Paris is poured to make it easier to lift the finished cast out of the ground. (The talcum powder can be omitted, although the soil will tend to stick to the Plaster of Paris as it is removed.)

6. If you don't have access to animal or bird tracks near school, your students can make casts of their own foot prints, or of other objects.

They will need a source of damp soil or sand, either in the classroom, or near the school. They should make an impression in the sand with a hand, foot, key, shell, or any other simple object that they choose. After a frame has been placed around the track and the track has been dusted with powder, the wet Plaster of Paris can be poured.

7. Prepare the supply table by first covering it with newspaper. If you are going to do this activity indoors, provide newspapers for your students to cover the area in which they will be stirring and pouring the Plaster of Paris. Lay out the cardboard frames, the stirring containers, and the Plaster of Paris.

8. If you are doing this activity outside, each group will have to carry a container of water in addition to a cardboard frame, a container of dry Plaster of Paris and a stirrer.

9. Discuss the procedure with your students before they add water to the Plaster of Paris. First they have to find a track, then fit the cardboard frame around it, and dust with powder. Then, and only then, should they add the water to the dry Plaster of Paris. (The tendency is to prepare too much Plaster of Paris too early.) One cup is sufficient for the size track that would be framed by cardboard from a 1 or 2 quart milk container. The Plaster of Paris should be poured to a depth of 1 – 2 inches (2 – 5 cm). Excess Plaster of Paris can be shared with other groups, or left in the containers until it hardens. The hardened material can be removed and discarded in the garbage.

10. The cast should be allowed to dry for about half an hour. It can then be lifted out of the soil gently, and carried back to class. The pieces of dirt that are attached to the hardened Plaster of Paris should not be removed until the cast is thoroughly dry (up to a week later). At that time the cast can be cleaned and labeled, or painted. A ruler should be provided to take measurements for Results and Observations question 3. (This step can be omitted, or it can be done at a different time during the year.)

11. If tracks are saved from year to year, an impressive collection can be accumulated that is very useful in many different measuring, graphing, and classifying activities.

TEACHER'S GUIDE
WHAT CAN WE DO WITH ALL OF OUR OLD PAPER?
Recycling

GOALS:

1. To demonstrate one way of making paper.
2. To describe something that can be done to reduce some of the paper waste.

SAFETY:

An apron should be worn. The mixer or blender should be operated by an adult, or under adult supervision.

PREPARATION AND HINTS:

1. It is best if each student prepares their own paper. If this is not possible, have students work in groups, and cut the final sheet of paper into pieces so that each group member can have a piece of "homemade" paper.

2. Obtain a mixer or blender, newspaper, and a large bowl.

3. Have each student tear one or two sheets of newspaper into very small pieces, and put all the pieces in the large bowl. Add enough water to cover all of the paper. Allow the paper to soak overnight. (To save time, use very hot water, and begin mixing as soon as all the paper has been added.)

4. Obtain screening to spread the paper pulp on. One piece for each student is the ideal, and the best size is about 12 inches square. (Smaller screens can be used with no problem, but the final size of the paper depends upon the size of the screen.)

5. Obtain several newspapers for each student, or group of students to use as they drain the water out of their paper pulp.

6. Your students will need a cup or ladle to pour the paper pulp onto the screen, and plastic wrap to cover the upper surface of their paper as it dries.

7. Prepare the supply table by first covering it with newspaper. Provide newspapers for your students to cover their work areas, and to drain the paper pulp. Place the screens, plastic wrap, and cups to scoop the paper pulp on the table.

8. Beat the wet pieces of paper with a mixer or blender. Add 3 – 6 tablespoons of cornstarch. (The starch gives the paper strength. You can make paper without starch, so the amount of starch that you add is not critical.) Continue to beat until the pulp is a smooth, pancake batter consistency. (Different mixers will produce different degrees of "smoothness" If your mixer doesn't seem to be making "pancake batter" don't worry about it, the pulp can still be used to make paper.)

9. Discuss the procedure with your students before they take the pulp (they will need about ½ cup of pulp) and pour it onto their screens. If they are each making their own paper, they should take turns holding the screen for each other as the pulp is poured on the screen.) Tell them where to put their paper to dry.

10. Allow time the next day to check on the paper. If the paper is not dry, the screen should be moved to dry newspaper for another day. When the paper is dry, it can be carefully peeled from the screen.

TEACHER'S GUIDE
GETTING TO KNOW YOU
Learning The Parts Of A Microscope

GOALS:

1. To carry a microscope properly.
2. To identify and correctly label certain parts of a microscope.
3. To describe the function of certain parts of a microscope.
4. To observe what part of the microscope moves when the coarse focus knob is turned.
5. To identify several factors that might prevent a lighted field of view from being seen when looking through the eyepiece.

SAFETY:

Always use two hands when carrying a microscope (one on the arm, and one under the base). If the microscope has a plug, be sure that the cord doesn't hang off the edge of the table. Never touch anyone who is looking through the objective of a microscope, as they might hurt their eye.

PREPARATION AND HINTS:

1. It is ideal if each student can use their own microscope, but in reality they can work in groups. The size of the groups is determined by the number of microscopes that are available. (Several companies sell microscopes that are less expensive that the models used on the high school level, that are designed for elementary students. Your high school science department will have biology catalogues that list these microscopes under the heading of microscopes – elementary. You might also be able to borrow microscopes from your high school.)

2. Arrange the microscopes neatly on the supply table, or on the cart that is used to transport them. Adjust each microscope so that the cord (if the microscope has one) is wrapped neatly around the base, the diaphragm is adjusted to the largest hole, and the low power (shortest) objective is in position. Provide crayons, or colored pencils.

3. Demonstrate the proper way to carry a microscope, then have your students carry a microscope back to their table.

4. Follow the procedures in the student writeup. Caution your students that they should never force any part of the microscope. (Students often try to turn the coarse focus knob past the point where the body tube ceases to move. This can damage the microscope, making it difficult or impossible to focus.)

5. Remind your students that they should return the microscopes in the condition that they found them in. The cord should be neatly wrapped around the base, the diaphragm should be adjusted to the large hole, and the low power objective should be in position.

6. You may be tempted to gloss over this activity because it seems so basic. Good microscope technique is learned by starting with the basics. The remaining activities in this book assume that your students have successfully completed this activity.

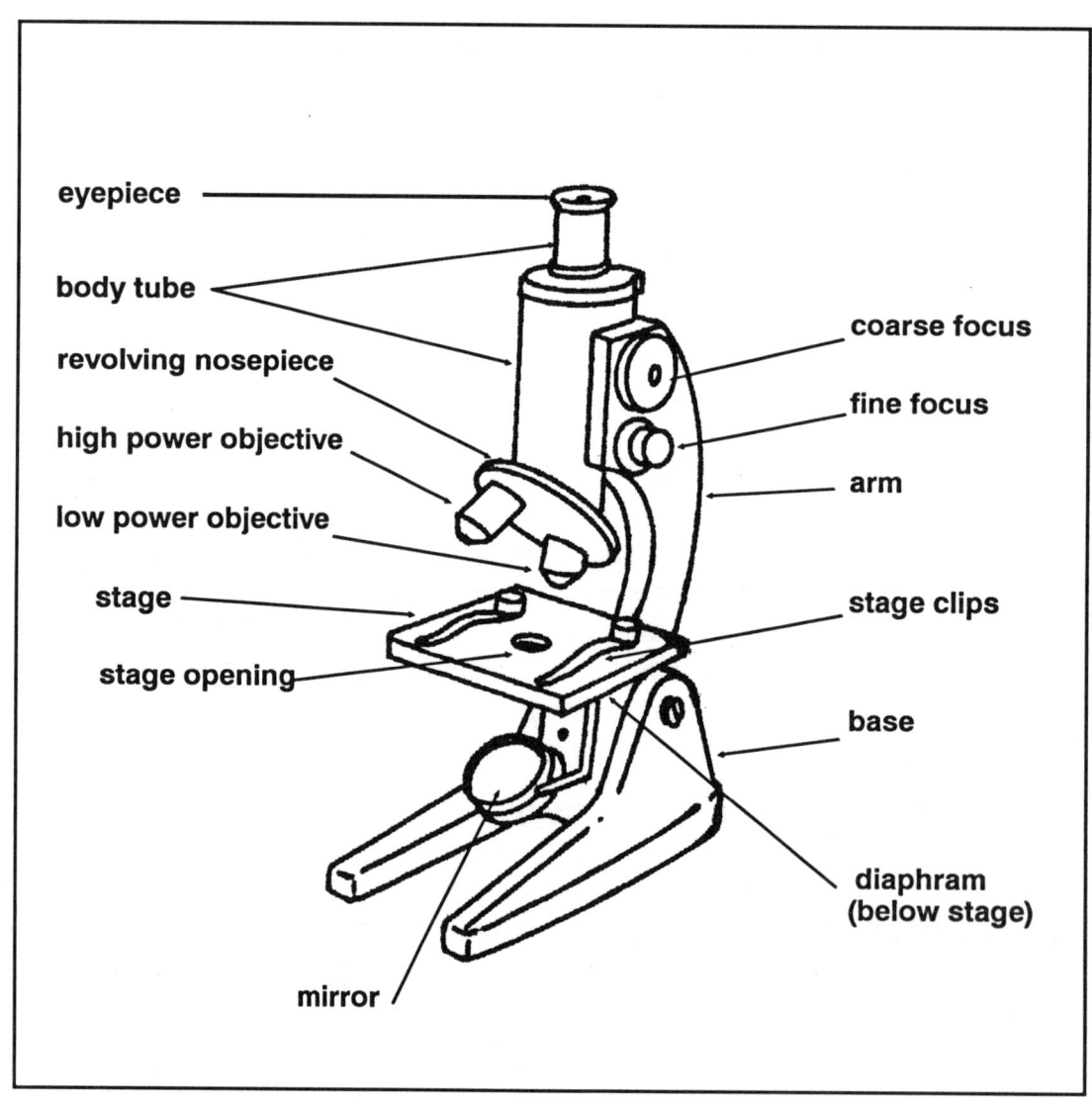

TEACHER'S GUIDE
UP CLOSE AND PERSONAL

Using A Microscope To Observe The Letter "v"

GOALS:

1. To practice carrying a microscope properly.
2. To prepare a wet mount.
3. To use the coarse focus knob to focus the microscope under low power.
4. To observe and sketch the magnified image of the letter "v".
5. To conclude that the microscope both magnifies and inverts the image.

SAFETY:

Always use two hands when carrying a microscope, one on the arm, and one under the base. If the microscope has a plug, be sure that the cord doesn't hang off the edge of the table. Never touch anyone who is looking through the objective of a microscope, as they might hurt their eye.

PREPARATION AND HINTS:

1. Your students should complete Activity 45 – *Getting To Know You*, before they begin this activity.

2. It is ideal if each student can use their own microscope, but in reality they can work in groups. The size of the groups is determined by the number of microscopes that are available.

3. Arrange the microscopes neatly on the supply table, or on the cart that is used to transport them. Check that each microscope has the cord wrapped neatly around the base, that the diaphragm is adjusted to the largest hole, and that the low power (shortest) objective is in position.

4. Obtain microscope slides and cover slips. Both are available in glass or plastic. Glass slides and cover slips can be cleaned and used repeatedly. Plastic slides and cover slips scratch easily, and must be discarded after several uses. A word of caution if you use glass cover slips. They are thin, and may break easily as your students try to

clean them. A good compromise is to use glass microscope slides, and plastic cover slips.

5. Your students will need water to prepare a wet mount. Each student or group, can use an eye dropper filled with water. The water can also be placed in small labeled squeeze bottles (used contact lens containers for example), and dispensed directly onto the slide.

6. A newspaper must be available as a source for the specimens. Your students will cut the lower case letter "v" from the paper. They should use letters that are a normal print size rather than those from headlines or classified advertisements. Let each group of students have a sheet of newspaper and a pair of scissors. They can then choose their own letter v and cut it out.

7. Arrange the microscope slides, cover slips, newspaper pages, scissors, eye droppers, and paper towels on the supply table.

8. Review the correct way to carry a microscope, and remind your students that they should never force any part of the microscope.

9. Demonstrate the correct technique for preparing a wet mount.

10. As they sketch the image of the letter v, remind your students that they should sketch exactly what they see. They should move the microscope slide on the stage to get as much of the image as possible within the field of view, but if the entire image does not fit within the field of view, they should only draw the part that they see. If the entire image does fit within the field of view, but it looks very small, the sketch should accurately represent the small size.

11. If your students complain that the image of the letter v is fuzzy or cloudy, a dirty lens may be the culprit. Lens paper (available in stores that sell photographic supplies) dampened with a drop of water and rubbed on the eyepiece and low power objective lenses is usually adequate to clean the lenses. If the lenses are exceptionally dirty, 1 teaspoon (5 mL) of ammonia can be added to 1 cup (236 mL) of water, and this mixture can be used to dampen the lens paper.

12. Remind your students that they should return the microscopes in the condition that they found them in – with the cord neatly wrapped around the base, the diaphragm adjusted to the largest hole, and the low power objective in position. Tell your students where to return the clean microscope slides, and whether they should clean their cover slips, or discard them.

TEACHER'S GUIDE
HOW CHEEKY ARE YOU?

Using A Microscope To Observe Stained Cheek Cells

GOALS:

1. To practice carrying a microscope properly.
2. To practice preparing a wet mount.
3. To prepare a stained wet mount.
4. To practice using the coarse focus knob to focus the microscope under low power.
5. To practice using the fine focus knob to focus the microscope under high power.
6. To observe and sketch the magnified image of stained cheek cells.
7. To conclude that animal cells contain a nucleus and cytoplasm.

SAFETY:

Always use two hands when carrying a microscope, one on the arm, and one under the base. If the microscope has a plug, be sure that the cord doesn't hang off the edge of the table. Never touch anyone who is looking through the objective of a microscope, as they might hurt their eye. Caution your students that the iodine stain is a poison, and that if it gets on their fingers or clothes, it will stain.

PREPARATION AND HINTS:

1. Your students should complete Activity 45 – *Getting To Know You*, before they begin this activity.
2. It is ideal if each student can use their own microscope, but in reality they can work in groups. The size of the groups is determined by the number of microscopes that are available.
3. Arrange the microscopes neatly on the supply table, or on the cart that is used to transport them. Check that each microscope has the cord wrapped neatly around the base, that the diaphragm is adjusted to the largest hole, and that the low power (shortest) objective is in position.

4. Obtain microscope slides and cover slips. Both are available in glass or plastic. Glass slides and cover slips can be cleaned and used repeatedly. Plastic slides and cover slips scratch easily, and must be discarded after several uses. A word of caution if you use glass cover slips. They are thin, and may break easily as your students try to clean them. A good compromise is to use glass microscope slides, and plastic cover slips.

5. Your students will need iodine stain to prepare the stained wet mount. If you have done earlier activities in which you used iodine solution to indicate the presence of starch, you may use the same iodine for this activity. If you have to prepare the iodine stain solution, mix 1 teaspoon (5 mL) of medicinal iodine with 1 cup (236 mL) of water. The iodine stain can be stored in small labeled plastic squeeze bottles (such as used contact lens solution containers), and dispensed directly from the container. If such continers are unavailable, the iodine stain can be placed in small labeled jars, beakers or cups, and dispensed with eye droppers.

6. Toothpicks will be needed to remove cheek cells. Flat toothpicks are easier to use than round toothpicks. Instruct your students to scrape saliva from the inside of their cheek. Some students have a tendency of scraping so hard that they remove enough cells to draw blood. This is not necessary. The inner cheek is always shedding cells, which are found in the saliva that is scraped off the cheek. These cells are not visible until they are stained and observed microscopically.

7. Arrange the microscope slides, cover slips, toothpicks, iodine stain, colored pencils or crayons, and paper towels on the supply table.

8. Review with your students the correct way to carry a microscope, and remind them that they should never force any part of the microscope.

9. Review the correct technique for preparing a wet mount.

10. Remind your students that they should sketch exactly what they see, after they adjust the slide so that they can see the image in the center of the field of view

11. Stress that the coarse focus knob is **not** to be used when the high power objective is in place.

12. If your students complain that the image of the cheek cells is fuzzy or cloudy, a dirty lens may be the culprit. Lens paper (available in store that sell photographic supplied) dampened with water, and rubbed on the eyepiece and objective lenses is usually adequate to clean the lenses. If the lenses are exceptionally dirty, a drop of a

solution of 1 teaspoon (5 mL) of ammonia and 1 cup (236 mL) water may be used to dampen the lens paper.

13. Remind your students that they should return the microscopes in the condition that they found them in — with the cord neatly wrapped around the base, the diaphragm adjusted to the largest hole, and the low power objective in position. Tell your students where to return the clean microscope slide, and whether they should clean their cover slips, or discard them.

TEACHER'S GUIDE
CAN YOU MAKE A BRICK WALL USING ONION CELLS?

Using A Microscope To Observe Stained Cells From The Membrane Of An Onion

GOALS:

1. To practice carrying a microscope properly.

2. To practice preparing a stained wet mount.

3. To practice using the coarse focus knob to focus the microscope under low power.

4. To practice using the fine focus knob to focus the microscope under high power.

5. To observe that the image of the specimen moves in one direction when the slide is moved in the opposite direction.

6. To observe and sketch the magnified image of stained onion membrane cells.

7. To conclude that a plant cell contains a nucleus and cytoplasm, and is surrounded by a cell wall.

SAFETY:

Always use two hands when carrying a microscope, one on the arm, and one under the base. If the microscope has a plug, be sure that the cord doesn't hang off the edge of the table. Never touch anyone who is looking through the objective of a microscope, as they might hurt their eye. Caution your students that the iodine stain is a poison, and that if it gets on their fingers or clothes, it will stain.

PREPARATION AND HINTS:

1. Your students should complete Activity 45 – *Getting To Know You*, before they begin this activity.

2. It is ideal if each student can use their own microscope, but in reality they can work in groups. The size of the groups is determined by the number of microscopes that are available.

3. Arrange the microscopes neatly on the supply table, or on the cart that is used to transport them. Check that each microscope has the cord wrapped neatly around the base, that the diaphragm is adjusted to the largest hole, and that the low power (shortest) objective is in position.

4. Obtain microscope slides and cover slips. Both are available in glass or plastic. Glass slides and cover slips can be cleaned and used repeatedly. Plastic slides and cover slips scratch easily, and must be discarded after several uses. A word of caution if you use glass cover slips. They are thin, and may break easily as your students try to clean them. A good compromise is to use glass microscope slides, and plastic cover slips.

5. Your students will need iodine stain to prepare the stained wet mounts. If you have done earlier labs in which you used iodine solution to indicate the presence of starch, you may use the same iodine for this activity. If you have to prepare the iodine stain solution, mix 1 teaspoon (5 mL) of medicinal iodine with 1 cup (236 mL) of water. The iodine stain can be stored in small labeled plastic squeeze bottles (such as used contact lens solution containers), and dispensed directly from the container. If such containers are unavailable, the iodine stain can be placed in small labeled jars, beakers or cups, and dispensed with eye droppers.

6. A raw onion will first have to be sliced into about 8 wedges. Between each layer in a wedge is a thin membrane. This membrane will be removed and stained by your students. The easiest way for them to remove the membrane from the piece of onion is to snap the onion piece in two, and peel the inner membrane away, using fingers, or forceps (tweezers). A toothpick is sometimes useful in separating an edge of the membrane from the onion, so that the membrane can be peeled off.

7. Arrange the microscope slides, cover slips, onion wedges, toothpicks, iodine stain, colored pencils or crayons, and paper towels on the supply table.

8. Review the correct way to carry a microscope. Remind your students that they should never force any part of the microscope.

9. Review the correct technique for preparing a wet mount.

10. Remind your students that they should sketch exactly what they see, after they adjust the slide so that the image of the stained onion membrane cells covers the entire field of view.

11. Stress that the coarse focus knob should not be used when the high power objective is in place.

12. If your students complain that the image of the onion membrane cells is fuzzy or cloudy, a dirty lens may be the problem. Lens paper (available in stores that sell photographic supplies) dampened with water, and rubbed on the eyepiece and objective lenses is usually adequate to clean the lenses. If the lenses are exceptionally dirty, a drop of a solution of 1 teaspoon (5 mL) of ammonia and 1 cup (236 mL) water may be used to dampen the lens paper.

13. Remind your students that they should return the microscopes in the condition that they found them in – with the cord neatly wrapped around the base, the diaphragm adjusted to the large hole, and the low power objective in position. Tell your students where to return the clean microscope slides, and whether they should clean their cover slips, or discard them.

Basic Biology Science Activities 101

TEACHER'S GUIDE
GREEN POLKA DOTS

Using A Microscope To Observe The Chloroplasts
In Green Plant Cells

GOALS:

1. To practice carrying a microscope properly.

2. To practice preparing a stained wet mount.

3. To practice using the coarse focus knob to focus the microscope under low power.

4. To practice using the fine focus knob to focus the microscope under high power.

5. To compare the number of plant cells that fit across the diameter of the low power field of view with the number of cells that fit across the diameter of the high power field of view.

6. To observe and sketch the magnified image of Elodea cells.

7. To conclude that while the images seen under high power magnification are larger than those seen under low power magnification, fewer cells can be seen in the high power field of view.

8. To conclude that a green plant cell contains a nucleus, cytoplasm, and chloroplasts, all of which are surrounded by a cell wall.

SAFETY:

Always use two hands when carrying a microscope, one on the arm, and one under the base. If the microscope has a plug, be sure that the cord doesn't hang off the edge of the table. Never touch anyone who is looking through the objective of a microscope, as they might hurt their eye.

PREPARATION AND HINTS:

1. Your students should complete Activity 45 – *Getting To Know You*, before they begin this activity.

2. It is ideal if each student can use their own microscope, but in reality they can work in groups. The size of the groups is determined by the number of microscopes that are available.

3. Arrange the microscopes neatly on the supply table, or on the cart that is used to transport them. Check that each microscope has the cord wrapped neatly around the base, that the diaphragm is adjusted to the largest hole, and that the low power (shortest) objective is in position.

4. Obtain microscope slides and cover slips. Both are available in glass or plastic. Glass slides and cover slips can be cleaned and used repeatedly. Plastic slides and cover slips scratch easily, and must be discarded after several uses. A word of caution if you use glass cover slips. They are thin, and may break easily as your students try to clean them. A good compromise is to use glass microscope slides, and plastic cover slips.

5. Obtain <u>Elodea</u>, <u>Anacharis</u> or any water plant that has translucent leaves. Water plants can be found in pet stores that sell fish, in fresh water ponds and streams, and at the shores of salt water bays. Transport the plants in a container that is large enough to hold both the plants and the water that the plants were found in.

6. You can maintain the water plants in their water for several days if you keep the container near a light source. (A window will provide natural lighting, while a lamp will supply adequate amounts of artificial light.) The best leaves to examine microscopically are those that are bright green, small, and near the tips of the branches. These are the youngest, newest leaves, and will be the easiest to observe.

7. Your students will need eye droppers, or small squeeze bottles containing water taken from the plant container, to prepare wet mounts of the green plant leaves. They should use the water that is in the plant container to prepare their wet mounts.

8. Arrange the microscope slides, cover slips, water plants in the container of water, colored pencils or crayons, and paper towels on the supply table.

9. Review the correct way to carry a microscope. Remind your students that they should never force any part of the microscope.

10. Review the correct technique for preparing a wet mount.

11. Remind your students that they should sketch exactly what they see in the field of view, after they have adjusted the slide so that the image of the green plant cells covers the entire field of view.

12. Stress that the coarse focus knob not be used when the high power objective is in place.

13. If your students complain that the image of the green plant cells is fuzzy or cloudy, a dirty lens may be the problem. Lens paper (available in stores that sell photographic supplies) dampened with water, and rubbed on the eyepiece and objective lenses is usually adequate to clean the lenses. If the lenses are exceptionally dirty, a drop of a solution of 1 teaspoon (5 mL) of ammonia and 1 cup (236 mL) water may be used to dampen the lens paper

14. Remind your students that they should return the microscopes in the condition that they found them in—with the cord neatly wrapped around the base, the diaphragm adjusted to the largest hole, and the low power objective in position. Tell your students where to return the clean microscope slides, and whether they should clean their cover slips, or discard them.

TEACHER'S GUIDE
WHO LIVES HERE?

Using A Microscope To Observe The Different Microorganisms That Live In Pond Water

GOALS:

1. To practice carrying a microscope properly.

2. To practice preparing a wet mount.

3. To practice using the coarse focus knob to focus the microscope under low power.

4. To practice using the fine focus knob to focus the microscope under high power.

5. To observe and sketch the magnified image of a microorganism from a pond water sample.

6. To conclude that there are a wide variety of microorganisms that live in pond water.

SAFETY:

Always use two hands when carrying a microscope, one on the arm, and one under the base. If the microscope has a plug, be sure that the cord doesn't hang off the edge of the table. Never touch anyone who is looking through the objective of a microscope, as they might hurt their eye.

PREPARATION AND HINTS:

1. Your students should complete Activity 45 – *Getting To Know You*, before they begin this activity.

2. It is ideal if each student can use their own microscope, but in reality they can work in groups. The size of the groups is determined by the number of microscopes that are available.

3. Arrange the microscopes neatly on the supply table, or on the cart that is used to transport them. Check that each microscope has the cord wrapped neatly around the base, that the diaphragm is adjusted to the largest hole, and that the low power (shortest) objective is in position.

4. Obtain microscope slides and cover slips. Both are available in glass or plastic. Glass slides and cover slips can be cleaned and used repeatedly. Plastic slides and cover slips scratch easily, and must be discarded after several uses. A word of caution if you use glass cover slips. They are thin, and may break easily as your students try to clean them. A good compromise is to use glass microscope slides, and plastic cover slips.

5. Your students will need eye droppers to remove a drop of pond water for observation. The eye droppers can be left on a paper towel next to the container of pond water.

6. Collect a pond water sample that includes several cups of water from the edge of a pond, as well as some mud and decaying vegetation from the bottom of the pond. The culture should be allowed to settle in your classroom (near a window) before it is examined by your students. If you can't collect pond water, you can make a pond water culture using rainwater, water from a river or stream, or bottled spring water. Add a handful of dry grass or hay (including the seed heads), and several tablespoons of soil to the water. Place the container near a window. After several days a scum will form on the surface, and the pond water culture will begin to smell. At this point it can be used as a source of microorganisms. The microorganisms in the pond water culture will continue to survive and reproduce over a period of several weeks. If your students enjoy observing the pond water microorganisms, you could have them observe the culture every week. They will observe ecological succession, as one form of microorganism dominates one week, then gradually diminishes in number as a different group of organisms, better suited to the conditions in the pond water culture, increase in number. This type of succession will continue for several weeks.

7. If time is limited, have your students observe one sample of pond water, rather than three. You can have them select a drop of water from any level of the culture (top, middle, or bottom) or specify that they all take a sample from a specific layer.

8. Arrange the microscope slides, cover slips, pond water culture, eye droppers, colored pencils or crayons, and paper towels on the supply table.

9. Review the correct way to carry a microscope. Remind your students that they should never force any part of the microscope.

10. Review the correct technique for preparing a wet mount. In this case, the drop of pond water serves as both specimen and water. A cover slip is placed directly over the drop of pond water.

11. Remind your students that they should sketch exactly what they see, after they adjust the slide so that the image of the microorganism fills the entire field of view, or is centered within the field of view.

12. Stress that the coarse focus knob not be used when the high power objective is in place.

13. If your students complain that the image of the microorganisms are fuzzy or cloudy, a dirty lens may be the problem. Lens paper (available in stores that sell photographic supplies) dampened with water, and rubbed on the eyepiece and objective lenses is usually adequate to clean the lenses. If the lenses are exceptionally dirty, a drop of a solution of 1 teaspoon (5 mL) of ammonia and 1 cup (236 mL) water may be used to dampen the lens paper.

14. Remind your students that they should return the microscopes in the condition that they found them in – with the cord neatly wrapped around the base, the diaphragm adjusted to the largest hole, and the low power objective in position. Tell your students where to return the clean microscope slides, and whether they should clean their cover slips, or discard them.

Basic Chemistry Science Activities

by

Dr. C. V. Krishnan

TEACHER'S SECTION
USING A BALANCE

GOALS AND OBJECTIVES:

1. To learn the use of a double pan or beam balance for measuring masses.
2. To learn to level the beam of a double pan balance.
3. To learn that two similar looking objects may not have the same mass.
4. To learn that the mass of an object depends on how much matter it contains in a given space.
5. To learn that the mass of an object depends on the nature of the matter it contains.
6. To learn that the mass of an object remains the same whether the object remains as one piece or is divided into several pieces
7. To learn that measurements give better information than guessing.

DISCUSSION BEFORE THE LABORATORY ACTIVITY:

It is very important to stress the need for wearing an apron and safety goggles.

Emphasize that no eating is allowed during hands on activity. The materials used in this laboratory activity are common household food items. However students should learn to treat these materials as laboratory items and no laboratory item should be eaten or tasted.

All balances are costly, and students should learn to handle them carefully. Also the balances are sensitive to small differences in masses, and it is important to hold the balance only at the base if it has to be moved.

Discuss the reasons why the masses of two apples of similar size may or may not be different and why only a balance can determine their true masses.

Discuss the reasons why the masses of an apple and an orange of similar size may be different.

Discuss the reasons why the masses of different materials contained in cups of the same size may be different.

Discuss the need for carefully separating the shells and the nuts without losing any matter, and introduce the idea of accuracy.

TEACHER'S SECTION
AIR IS MATTER

GOALS AND OBJECTIVES:

1. To determine that air is matter and that air occupies space.
2. To learn that air occupies space even though one cannot see it.
3. To learn that no two substances can occupy the same space at the same time.

DISCUSSION BEFORE THE LABORATORY ACTIVITY:

It is very important to stress the need to wear an apron and safety goggles.

For safety purpose, do not let the students cut the neck of the plastic bottle.

Discuss the various properties of air and discuss ways of determining its properties.

Discuss the properties of the three phases of matter: solid, liquid, and gas.

TEACHER'S SECTION
HELIUM BALLOON

GOALS AND OBJECTIVES:

1. To show that air has mass.
2. To learn that more air will weigh more.
3. To show that helium is lighter than air and that you cannot weigh helium balloons easily.
4. To learn that only materials heavier than air can be weighed easily.

DISCUSSION BEFORE THE LABORATORY ACTIVITY:

It is very important to stress the need for wearing an apron and safety goggles.

Discuss the meaning of the words light and heavy or dense.

Discuss why substances such as wood and paper float in water whereas substances such as an iron nail and a coin sink.

Discuss why a person feels lighter during floating or swimming in water.

Discuss buoyancy.

Challenge students to name gases that are lighter than air. Natural gas (methane), helium, neon, hydrogen, and acetylene (ethyne, used for welding) are examples of gases that are lighter than air.

Challenge students to name gases that are heavier than air. Pure carbon dioxide, argon, nitrogen dioxide, and sulfur dioxide are examples of gases that are denser than air.

TEACHER'S SECTION
DRY ICE

GOALS AND OBJECTIVES:

1. To learn that dry ice is solid carbon dioxide.
2. To learn that dry ice is very cold.
3. To learn that dry ice does not wet the container.
4. To learn that dry ice can extinguish flames.
5. To learn that dry ice does not melt. Instead it sublimes.
6. To learn that when dry ice becomes a gas, it expands considerably.

DISCUSSION BEFORE THE LABORATORY ACTIVITY:

It is very important to stress the need for wearing an apron and safety goggles.

Challenge students to name substances that are very cold.

Discuss the properties of solids, liquids, and gases. Discuss phase changes. How do we get a liquid from a solid and a gas from a liquid? How do we get a liquid from a gas and a solid from a liquid?

Challenge students to say whether it is possible to change a solid into a gas without melting and whether it is possible to change a gas into a solid without condensing it into a liquid.

The melting point of regular ice is 0°C. Dry ice sublimes at −78.5°C at normal atmospheric pressure of 1 atmosphere. It is possible to melt dry ice into liquid carbon dioxide at temperatures greater than −57°C and when the pressure is greater than 5.2 atmospheres.

TEACHER'S SECTION
VOLUME OF A SOLID AND A GAS

GOALS AND OBJECTIVES:

1. To show, by using the same substance as an example, that a gas occupies much more space than a solid.

2. To find the volume occupied by 1 gram of carbon dioxide gas at room temperature.

DISCUSSION BEFORE THE LABORATORY ACTIVITY:

It is very important to stress the need for wearing an apron and safety goggles.

Ask students to name substances that are very cold.

Discuss the properties of solids, liquids, and gases.

Discuss phase changes. How do we get a liquid from a solid and a gas from a liquid? How do we get a liquid from a gas and a solid from a liquid?

Ask the students if it is possible to change a solid into a gas without melting it and if it is possible to change a gas into a solid without condensing it into a liquid.

In a solid, the molecules are very close to each other because of greater attraction among them. In a gas, the molecules are far apart and are in random motion. The molecules of a gas have little attraction among them. The same amount of solid, when changed into a gas, occupies a much larger volume than when it is a solid. One gram of carbon dioxide gas occupies a volume of about 555 milliliters at 25°Celsius and air pressure of one atmosphere, whereas one gram of carbon dioxide solid (dry ice) occupies a volume of only about 1 milliliter.

The melting point of regular ice is 0°C. Dry ice sublimes at –78.5°C at normal atmospheric pressure of 1 atmosphere. It is possible to melt dry ice into liquid carbon dioxide at temperatures greater than –57°C and when the pressure is greater than 5.2 atmospheres.

A solid occupies the least volume compared to the liquid and the gas produced by the same amount of solid. Water is an exception. Ice occupies more volume than liquid water produced from the same amount of ice.

It is not easy to find other substances that show this large volume change when they change from a solid into a gas.

TEACHER'S SECTION
VOLUME OF A SOLID

GOALS AND OBJECTIVES:

1. To determine the volumes of solids that have no regular shape.

2. To learn that the volume of an object is the same as the volume of water displaced by the object.

3. To learn that the volume of a large object is the same as the sum of the volumes of the several pieces obtained by dividing the large object into several pieces.

DISCUSSION BEFORE THE LABORATORY ACTIVITY:

It is very important to stress the need to wear an apron and safety goggles.

Discuss the meaning of volume of an object.

Show several objects to students and ask them to predict the comparative volumes of the objects. Ask them to suggest ways of checking out their predictions.

Discuss ways of measuring the volume of a large object. Ask them how to measure the volume of a large object using only a small graduated cylinder or baby nurser which has graduations in metric units.

Discuss the advantages of the water displacement technique for measuring volumes. Ask the students to design an apparatus for measuring the volume of a large object using the water displacement technique.

Only the volumes of solids that are not porous and that do not dissolve in water can be measured by water displacement.

Discuss the accuracies of measuring volumes using graduated cylinders of different sizes, graduated beakers, and baby nursing bottle. There are metric measurements on the baby nurser that can be conveniently used for measuring volumes.

TEACHER'S SECTION
VOLUME OF A SOLID WITH A REGULAR SHAPE

GOALS AND OBJECTIVES:

1. To determine the volumes of solids that have regular shapes.

2. To learn that the volume of an object is the same as the volume of water displaced by the object.

3. To learn that the volumes of objects that have regular shapes can be computed using mathematical formulas.

4. To learn that the volume of an object determined by the water displacement technique is the same as the volume determined by using an appropriate mathematical formula.

DISCUSSION BEFORE THE LABORATORY ACTIVITY:

It is important to stress the need for wearing an apron and safety goggles.

Discuss the meaning of volume of an object.

Show several objects to students and ask them to predict their comparative volumes. Ask them to suggest ways of checking their predictions.

Introduce the concept of the area of a square and a rectangle. Then introduce the concept of the volume of a cube and a rectangular solid.

Discuss the advantages of the water displacement technique for measuring volumes. Ask the students to design an apparatus for measuring the volume of a large object using water displacement technique.

Only volumes of solids that are not porous and that do not dissolve in water can be measured by water displacement.

Discuss the accuracies of measuring volumes using graduated cylinders of different sizes, graduated beakers, and baby nursing bottle. The metric measurements on the baby bottle can be conveniently used for measuring volumes.

Let the students practice cutting exact cubes and rectangular solids using substances such as apples and potatoes. This takes patience and extreme care. Cubes and rectangular solids of wood, plastic, and metals are also commercially available.

TEACHER'S SECTION
MOLECULES OF GASES

GOALS AND OBJECTIVES:

1. To determine that gas molecules are moving very fast in all directions.

2. To find out whether it is possible to determine the relative speeds of different molecules.

3. To find out whether it is possible to time the movement of molecules within short distances.

DISCUSSION BEFORE THE LABORATORY ACTIVITY:

It is very important to stress the need to wear an apron and safety goggles.

Discuss the relative motion of molecules in a solid, a liquid, and a gas.

Challenge the students to suggest ways to find out how molecules move in a solid, in a liquid and in a gas.

Ask them to predict the speed of gas molecules.

Discuss the term speed and the units for expressing speed.

Gas molecules move very fast. For example, the speed of nitrogen molecules at 25°C is about 515 meters per second. The speed of a minimum hurricane wind is 75 miles per hour or about 34 meters per second. When molecules are moving, they collide with each other. A molecule collides with other molecules about 5 billion times a second. The average distance traveled by a molecule between collisions is extremely small, about 200 times the diameter of the molecule. Bigger molecules move slower than smaller ones.

TEACHER'S SECTION
ESCAPING OF GAS MOLECULES

GOALS AND OBJECTIVES:

1. To determine whether gas molecules can escape from their containers.

2. To find out whether molecules of perfume escape more easily from a balloon or from a paper towel bag.

DISCUSSION BEFORE THE LABORATORY ACTIVITY:

It is very important to stress the need for wearing an apron and safety goggles.

Discuss with the students the relative motion of molecules in a solid, liquid, and gas.

Ask them to suggest ways to determine molecular motion in a solid, in a liquid, and in a gas.

Ask the students to predict the speed of gas molecules.

Ask them to predict the best way to store perfume and how to check out this prediction.

Ask them to explain how the smell of perfume comes out when a bottle of perfume is opened.

Have them suggest an experiment using water to find out whether the tiny holes in a rubber balloon are bigger or smaller than the tiny holes in a paper towel. (The water should remain in the balloon and the water should drip through the paper towel.)

Gas molecules move very fast. For example, the speed of nitrogen molecules at 25°C is about 515 meters per second. The speed of a minimum hurricane wind is 75 miles per hour or about 34 meters per second. When molecules are moving, they collide with each other. A molecule collides with other molecules about 5 billion times a second. The average distance traveled by a molecule between collisions is extremely small, about 200 times the diameter of the molecule. Bigger molecules move slower than smaller ones.

TEACHER'S SECTION
TEMPERATURE MEASUREMENTS USING AN ALCOHOL THERMOMETER AND A GAME THERMOMETER

GOALS AND OBJECTIVES:

1. To learn the use of an alcohol thermometer.
2. To learn the use of a game thermometer.
3. To learn the principle of an alcohol thermometer.
4. To learn the principle of a game thermometer.

DISCUSSION BEFORE THE LABORATORY ACTIVITY:

It is very important to stress the need to wear an apron and safety goggles.

Discuss heat energy and the flow of heat energy.

Discuss the meaning of temperature.

Temperature is a measure of the average kinetic energy of the molecules. Kinetic energy is the energy from the motion of molecules. There are three common temperature scales, Fahrenheit (F), Celsius (C), and Kelvin (K). The fixed points on the Fahrenheit and Celsius thermometers are based on the melting point of ice and the boiling point of water. It is assumed that ice melts at 32°F or 0°C and water boils at 212°F or 100°C at one atmospheric pressure. The more meaningful Kelvin scale is based on the lowest temperature possible to attain. The lowest possible temperature is −273°C. This temperature is called absolute zero or zero Kelvin. Scientists have succeeded in reaching very close to absolute zero but not absolute zero itself. According to this Kelvin scale (also called absolute scale), ice melts at 273 K and water boils at 373 K.

Discuss the basis of mercury and alcohol thermometers.

Discuss the fixed points of Fahrenheit and Celsius thermometers.

Discuss liquid crystal thermometers.

This experiment can also be carried out using a mercury thermometer instead of an alcohol thermometer. Mercury is toxic. In case of breakage, the spilled mercury should be carefully collected using special mercury spill kits and disposed according to school and state regulations.

The mercury and alcohol thermometers are based on the expansion or contraction of liquids with change in temperature. It is assumed that the liquid column rises in proportion to an increase in temperature.

The game thermometer is based on the expansion of air with increases in temperature. Increasing temperature will increase the pressure of a fixed volume of air. This increased pressure will force water to move up a tube until the outside and inside pressure are equal.

TEACHER'S SECTION
AIR PRESSURE

GOALS AND OBJECTIVES:

1. To show that air exerts pressure.

2. To show that it is extremely difficult to pull apart two metal hemispheres joined together by outside air pressure and inside vacuum.

3. To show that it is difficult to pull apart rubber suction cups due to the outside atmospheric pressure.

4. To show that water remains inside a narrow tube closed at one end because of outside air pressure.

DISCUSSION BEFORE THE LABORATORY ACTIVITY:

It is very important to stress the need for wearing an apron and safety goggles.

It is very important to stress the need to try to carefully pull apart the two rubber suction cups.

It is very important to stress the need to try to carefully pull apart the two metal hemispheres without getting hurt. Special care must be taken so as not to hit your face with the exhaust valve connector on the metal hemisphere.

As everybody knows, glass is very fragile, and the tube must be handled carefully.

Discuss the meaning of air pressure. The pressure of a gas is due to collisions of molecules on the walls of the container. The pressure increases when the number of molecules increases and when the temperature increases. The normal air pressure at sea level is one atmosphere (atm). Pressure is defined as the force per unit area. The International System (SI) unit of pressure is Pascals (Pa). The Pascal is defined as one newton per square meter where newton is the unit of force. One newton is the force needed to accelerate a mass of one kilogram by 1 meter per second per second. Another unit of pressure is torr, named after the Italian physicist, Evangelista Torricelli (1608 – 1647), who conducted several important experiments related to air pressure. Air pressure is also expressed in terms of the height of the mercury column inside a tube supported by outside air pressure (barometer).

$$1 \text{ atm} = 101330 \text{ Pa} = 760 \text{ torr} = 760 \text{ millimeters of mercury}$$

Gas molecules move very fast. For example, the speed of nitrogen molecules at 25°C is about 515 meters per second. The speed of a minimum hurricane wind is 75 miles per hour or about 34 meters per second. When molecules are moving, they collide with each other. A molecule collides with other molecules about 5 billion times a second. The average distance traveled by a molecule between collisions is extremely small, about 200 times the diameter of the molecule. Bigger molecules move slower than smaller ones.

TEACHER'S SECTION
MAGIC WITH A BALLOON AND TWO CUPS

GOALS AND OBJECTIVES:

1. To create a suction effect by rapidly inflating a balloon.
2. To find ways to create a partial vacuum.
3. To find the everyday practical uses for creating a partial vacuum.
4. To show that suction effect can be produced either by creating a partial vacuum or by increasing pressure on one side compared to the other.

DISCUSSION BEFORE THE LABORATORY ACTIVITY:

It is very important to stress the need for wearing an apron and safety goggles.

Discuss the meaning of suction.

Discuss the meaning of air pressure. The pressure of a gas is due to collisions of molecules on the walls of the container. The pressure increases when the number of molecules increases and when the temperature increases. The normal air pressure at sea level is one atmosphere (atm). Pressure is defined as the force per unit area. The International System (SI) unit of pressure is Pascals (Pa). The Pascal is defined as one newton per square meter where newton is the unit of force. One newton is the force needed to accelerate a mass of one kilogram by 1 meter per second per second. Another unit of pressure is torr, named after the Italian physicist, Evangelista Torricelli (1608 – 1647), who conducted several important experiments related to air pressure. Air pressure is also expressed in terms of the height of the mercury column inside a tube supported by outside air pressure (barometer).

$$1 \text{ atm} = 101330 \text{ Pa} = 760 \text{ torr} = 760 \text{ millimeters of mercury}$$

Discuss the meaning of vacuum.

Ask students to give examples where suction is used in everyday life.

TEACHER'S SECTION
MAGIC WITH PRESSURE OF AIR IN BALLOONS

GOALS AND OBJECTIVES:

1. To determine which balloon has more air pressure.
2. To show that the air pressure inside a greatly inflated balloon is less than the air pressure inside a less inflated balloon.
3. To show that it is more difficult to stretch a balloon at the beginning.

DISCUSSION BEFORE THE LABORATORY ACTIVITY:

It is very important to stress the need for wearing an apron and safety goggles.

Discuss the meaning of air pressure. The pressure of a gas is due to collisions of molecules on the walls of the container. The pressure increases when the number of molecules increases and when the temperature increases. The normal air pressure at sea level is one atmosphere (atm). Pressure is defined as the force per unit area. The International System (SI) unit of pressure is Pascals (Pa). The Pascal is defined as one newton per square meter where newton is the unit of force. One newton is the force needed to accelerate a mass of one kilogram by 1 meter per second per second. Another unit of pressure is torr, named after the Italian physicist, Evangelista Torricelli (1608 – 1647), who conducted several important experiments related to air pressure. Air pressure is also expressed in terms of the height of the mercury column inside a tube supported by outside air pressure (barometer).

$$1 \text{ atm} = 101330 \text{ Pa} = 760 \text{ torr} = 760 \text{ millimeters of mercury}$$

Why does the air come out of an inflated untied balloon?

Is it easier or more difficult to inflate a balloon at the beginning?

Gas molecules move very fast. For example, the speed of nitrogen molecules at 25°C is about 515 meters per second. The speed of a minimum hurricane wind is 75 miles per hour or about 34 meters per second. When molecules are moving, they collide with each other. A molecule collides with other molecules about 5 billion times a second. The average distance traveled by a molecule between collisions is extremely small, about 200 times the diameter of the molecule. Bigger molecules move slower than smaller ones.

Ask the students to predict whether the air pressure inside a balloon inflated fully is greater or less than the pressure of air inside a balloon only partly. Ask them to suggest ways to check their predictions.

TEACHER'S SECTION
AIR EXERTS PRESSURE

GOALS AND OBJECTIVES:

1. To determine that air is matter and that air occupies space.
2. To learn that air exerts pressure.
3. To learn that no two substances can occupy the same space at the same time.

DISCUSSION BEFORE THE LABORATORY ACTIVITY:

It is very important to stress the need for wearing an apron and safety goggles.

Discuss the various properties of air and discuss ways of determining its properties.

Discuss the properties of the three phases of matter, solid, liquid, and gas.

Ask the students to push a balloon that is slightly larger than the mouth of a bottle into the bottle. Why is that difficult to do?

Discuss the meaning of air pressure. The pressure of a gas is a result of collisions of molecules with the walls of the container. The pressure increases with the number of molecules and with the temperature. Normal air pressure at sea level is one atmosphere (atm). Pressure is defined as the force per unit area. The SI unit of pressure is pascals (Pa). The pascal is defined as one newton per squaremeter where newton is the unit of force. One newton is the force needed to accelerate a mass of one kilogram by 1 meter per second per second. Another unit of pressure is torr, named after the Italian physicist, Evangelista Torricelli (1608–1647), who conducted several important experiments related to air pressure. Air pressure is also expressed in terms of the height of the mercury column inside a tube supported by outside air pressure (a barometer).

$$1 \text{ atm} = 101330 \text{ Pa} = 760 \text{ torr} = 760 \text{ millimeters of mercury}$$

Gas molecules move very fast. For example, the speed of nitrogen molecules at 25°C is about 515 meters per second. The speed of a minimum hurricane wind is 75 miles per hour or about 34 meters per second. When molecules are moving, they collide with each other. A molecule collides with other molecules about 5 billion times a second. The average distance traveled by a molecule between collisions is extremely small, about 200 times the diameter of the molecule. Bigger molecules move slower than smaller ones.

TEACHER'S SECTION
FUN WITH AIR

GOALS AND OBJECTIVES:

1. To have fun with some commercially available and easily made toys using the properties of air.

2. To determine that air is matter and that air occupies space.

3. To learn that no two substances can occupy the same space at the same time.

4. To learn that the principle of suction is used in some toys.

DISCUSSION BEFORE THE LABORATORY ACTIVITY:

It is very important to stress the need to wear an apron and safety goggles.

Discuss the various properties of air and the ways to determine its properties.

Discuss the properties of the three phases of matter: solid, liquid, and gas.

Gas molecules move very fast. For example, the speed of nitrogen molecules at 25°C is about 515 meters per second. The speed of a minimum hurricane wind is 75 miles per hour or about 34 meters per second. When molecules are moving, they collide with each other. A molecule collides with other molecules about 5 billion times a second. The average distance traveled by a molecule between collisions is extremely small, about 200 times the diameter of the molecule. Bigger molecules move slower than smaller ones.

Discuss the meaning of air pressure. The pressure of a gas is due to collisions of molecules on the walls of the container. The pressure increases when the number of molecules increases and when the temperature increases. The normal air pressure at sea level is one atmosphere (atm). Pressure is defined as the force per unit area. The International System (SI) unit of pressure is Pascals (Pa). The Pascal is defined as one newton per square meter where newton is the unit of force. One newton is the force needed to accelerate a mass of one kilogram by 1 meter per second per second. Another unit of pressure is torr, named after the Italian physicist, Evangelista Torricelli (1608 – 1647), who conducted several important experiments related to air pressure. Air pressure is also expressed in terms of the height of the mercury column inside a tube supported by outside air pressure (barometer).

1 atm = 101330 Pa = 760 torr = 760 millimeters of mercury

Discuss the meaning of suction.

Ask the students to design and construct a toy based on the principles of air pressure, suction or vacuum, and the properties of air.

Ask the students to push a balloon that is slightly bigger than the mouth of a bottle into the bottle. Discuss the reasons for the difficulty in accomplishing this.

It is extremely difficult to get the juice to flow from a can by punching only one hole. In order for the juice to flow out, air must be able to enter. It is difficult for the air to enter and the juice to flow out at the same time through one hole. The juice comes out easily if another hole is punched to allow the air to enter.

Air pressure can support a column of water in a tube about 10.3 meters high and closed at one end. To show this suspension of water in a tube in air effectively, the mouth of the tube must be narrow and uniform. The air pressure supporting a high column of water in any size tube can be shown effectively by keeping the open end of the tube under the water in a tray.

TEACHER'S SECTION
MAGIC BY SQUEEZING AIR

GOALS AND OBJECTIVES:

1. To produce a water fountain using increased air pressure.
2. To learn that air occupies space and exerts pressure.
3. To learn that it is not possible to blow air into a bottle for more than a few seconds.

DISCUSSION BEFORE THE LABORATORY ACTIVITY:

It is very important to stress the need for wearing an apron and safety goggles.

Discuss the various properties of air and discuss ways of determining those properties.

Ask the students to push a balloon slightly bigger than the mouth of a bottle into the bottle. Discuss the reasons why it is difficult to do this.

Ask the students to design a way to blow the maximum amount of air into a bottle and then to measure this air.

Discuss the meaning of air pressure. The pressure of a gas is due to collisions of molecules on the walls of the container. The pressure increases when the number of molecules increases and when the temperature increases. The normal air pressure at sea level is one atmosphere (atm). Pressure is defined as the force per unit area. The International System (SI) unit of pressure is Pascals (Pa). The Pascal is defined as one newton per square meter where newton is the unit of force. One newton is the force needed to accelerate a mass of one kilogram by 1 meter per second per second. Another unit of pressure is torr, named after the Italian physicist, Evangelista Torricelli (1608 – 1647), who conducted several important experiments related to air pressure. Air pressure is also expressed in terms of the height of the mercury column inside a tube supported by outside air pressure (barometer).

$$1 \text{ atm} = 101330 \text{ Pa} = 760 \text{ torr} = 760 \text{ millimeters of mercury}$$

TEACHER'S SECTION
COLD AIR AND HOT AIR

GOALS AND OBJECTIVES:

1. To determine whether cold air or hot air occupies more space.
2. To learn that a gas contracts on cooling and expands on heating.
3. To learn that when a gas is heated, the number of collisions against the walls increases and the pressure increases.

DISCUSSION BEFORE THE LABORATORY ACTIVITY:

It is very important to stress the need for wearing an apron and safety goggles.

Discuss the need for wearing insulated waterproof gloves.

It is very important to learn to use the hot water very carefully.

Discuss the various properties of air and discuss ways of determining its properties.

Discuss the expansion of solids, liquids, and gases on heating.

Gas molecules move very fast. For example, the speed of nitrogen molecules at 25°C is about 515 meters per second. The speed of a minimum hurricane wind is 75 miles per hour or about 34 meters per second. When molecules are moving, they collide with each other. A molecule collides with other molecules about 5 billion times a second. The average distance traveled by a molecule between collisions is extremely small, about 200 times the diameter of the molecule. Bigger molecules move slower than smaller ones.

Discuss the meaning of air pressure. The pressure of a gas is due to collisions of molecules on the walls of the container. The pressure increases when the number of molecules increases and when the temperature increases. The normal air pressure at sea level is one atmosphere (atm). Pressure is defined as the force per unit area. The International System (SI) unit of pressure is Pascals (Pa). The Pascal is defined as one newton per square meter where newton is the unit of force. One newton is the force needed to accelerate a mass of one kilogram by 1 meter per second per second. Another unit of pressure is torr, named after the Italian physicist, Evangelista Torricelli (1608 – 1647), who conducted several important experiments related to air pressure. Air pressure is also expressed in terms of the height of the mercury column inside a tube supported by outside air pressure (barometer).

1 atm = 101330 Pa = 760 torr = 760 millimeters of mercury

Molecules are closer to each other when they are cold. And they occupy less volume when they are cold. The same is true for solids, liquids, and gases. Thus a gas occupies the most volume and a solid occupies the least volume. There are some exceptions. For example, ice occupies more space than liquid water.

TEACHER'S SECTION
MOISTURE IN AIR

GOALS AND OBJECTIVES:

1. To test for moisture in air.
2. To learn that moisture is present in air.

DISCUSSION BEFORE THE LABORATORY ACTIVITY:

It is very important to stress the need for wearing an apron and safety goggles.

Discuss how moisture is produced in the air.

Discuss humidity and weather.

Discuss how frost is formed.

Discuss the three phases of matter.

Discuss the meaning of steam, water vapor, moisture and humidity.

Discuss ways to produce water and steam from ice and to produce water and ice from steam.

Ice melts at 0°C(elsius) and water freezes at 0°C. Water boils at 100°C and steam condenses at 100°C. Moisture can condense on any surface that is colder than the room temperature. The actual condensation depends on the amount of moisture present in the air and the temperature of the cold surface.

Ask students to suggest a method for detecting the presence of moisture in the air.

Basic Chemistry Science Activities 133

TEACHER'S SECTION
PHYSICAL CHANGES

GOALS AND OBJECTIVES:

1. To learn that the melting of a solid is a physical change.
2. To learn that the freezing of a liquid is a physical change.
3. To learn that the expansion of a liquid is a physical change.
4. To learn that a physical change does not change the composition of a substance.

DISCUSSION BEFORE THE LABORATORY ACTIVITY:

It is important to stress the need for wearing an apron and safety goggles.

Caution students to handle cups containing boiling water with extreme care.

If possible use an alcohol thermometer instead of a mercury thermometer.

Discuss the properties of matter.

Discuss phase changes, solid to liquid, liquid to gas, gas to liquid, and liquid to solid.

Discuss expansion and contraction of objects due to temperature changes.

Temperature is a measure of the average kinetic energy of the molecules. Kinetic energy is the energy from the motion of molecules. There are three common temperature scales, Fahrenheit (F), Celsius (C), and Kelvin (K). The fixed points on the Fahrenheit and Celsius thermometers are based on the melting point of ice and the boiling point of water. It is assumed that ice melts at 32°F or 0°C and water boils at 212°F or 100°C at one atmospheric pressure. The more meaningful Kelvin scale is based on the lowest temperature possible to attain. The lowest possible temperature is –273°C. This temperature is called absolute zero or zero Kelvin. Scientists have succeeded in reaching very close to absolute zero but not absolute zero itself. According to this Kelvin scale (also called absolute scale), ice melts at 273 K and water boils at 373 K.

Remind students that ice, liquid water, and steam are all different forms of the same substance, water.

Discuss physical properties and chemical properties. Ask students to give some examples of both of these properties.

Remind students that a physical change does not change the composition of the substance.

TEACHER'S SECTION
MELTING POINT OF ICE

GOALS AND OBJECTIVES:

1. To find the nature of temperature change during the phase change, solid to liquid.
2. To measure the melting point of ice.
3. To learn that the melting of a solid is a physical property.
4. To learn to represent data on a graph.

DISCUSSION BEFORE THE LABORATORY ACTIVITY:

It is very important to stress the need for wearing an apron and safety goggles.

Discuss the properties of matter.

Discuss phase changes, solid to liquid, liquid to gas, gas to liquid, and liquid to solid.

Remind students that ice, liquid water, and steam are all different forms of the same substance, water.

Discuss physical properties and chemical properties. Ask students to give some examples of both of these properties.

Remind students that a physical change does not change the composition of the substance.

Ask students to name solid substances that melt rather easily. Compared to candle wax, butter, and margarine, sugar has a higher melting point. Also sugar decomposes rather easily. If it decomposes, that will be a chemical change. Melting is a physical change. Substances such as common salt, baking soda, and washing soda have high melting points.

When a solid is heated uniformly, the temperature should increase uniformly. However, the temperature does not increase indefinitely. The temperature will increase uniformly until the melting temperature of the solid is reached. At the melting temperature or melting point, the temperature remains the same even if the substance is continuously heated. This is true for all crystalline substances. At the melting point, heat energy changes form and is stored as potential energy.

A thermometer cannot measure potential energy. It can only measure energy caused by the motion of molecules. Temperature is a measure of the average kinetic energy of the molecules. Even though one expects, from common sense, that the temperature of a substance will increase continuously on continuous heating, this does not happen at the melting point.

Kinetic energy is the energy from the motion of molecules. There are three common temperature scales, Fahrenheit (F), Celsius (C), and Kelvin (K). The fixed points on the Fahrenheit and Celsius thermometers are based on the melting point of ice and the boiling point of water. It is assumed that ice melts at 32°F or 0°C and water boils at 212°F or 100°C at one atmospheric pressure. The more meaningful Kelvin scale is based on the lowest temperature possible to attain. The lowest possible temperature is –273°C. This temperature is called absolute zero or zero Kelvin. Scientists have succeeded in reaching very close to absolute zero but not absolute zero itself. According to this Kelvin scale (also called absolute scale), ice melts at 273 K and water boils at 373 K.

Sharp melting points are obtained only for crystalline substances. Substances such as candle wax and butter are not crystalline and therefore it is not possible to get sharp melting points.

Introduce students to the technique of drawing graphs and the advantages of graphs in interpreting and understanding results. Introduce the names, ordinate and abscissa, for the Y and X coordinates. Discuss the importance of labeling the axes and writing the units. They should learn to indicate data points as small circles or small squares instead of points, so that the data points can be easily recognized. It is extremely important for them to realize that a straight line graph need not touch all the data points. The points may be spread equally on both the sides of the line so that the best line is obtained. This is an important point and should be stressed many times.

Basic Chemistry Science Activities 137

TEACHER'S SECTION
FREEZING POINT OF WATER

GOALS AND OBJECTIVES:

1. To examine the nature of temperature change during the phase change, liquid to solid.
2. To measure the freezing point of water.
3. To learn that the freezing of a liquid is a physical property.
4. To learn that the freezing of a liquid and the melting of a solid obtained from the same substance take place at the same temperature.
5. To learn to represent data in a graph.

DISCUSSION BEFORE THE LABORATORY ACTIVITY:

It is important to stress the need to wear an apron and safety goggles.

Discuss the properties of matter.

Discuss phase changes, solid to liquid, liquid to gas, gas to liquid, and liquid to solid. Introduce the fact that the melting and freezing of a substance take place at the same temperature for crystalline substances.

Remind students that ice, liquid water, and steam are all different forms of the same substance, water.

Discuss physical properties and chemical properties. Ask students to give some examples of both of these properties.

Remind students that a physical change does not change the composition of the substance.

Ask students to name liquid substances that will freeze easily. Remind them of the freezing of water in pipes during winter.

Vegetable oils and baby oils also freeze easily. Molten candle wax, molten butter and molten margarine also freeze easily.

When a liquid cools uniformly, the temperature should decrease uniformly. However, the temperature does not decrease indefinitely. The temperature will decrease uniformly until the freezing temperature of the liquid is reached. At the freezing temperature or freezing point, the temperature remains the same even if the substance is continuously cooled. This is true for all substances that will give solid crystals. At the freezing point, the stored potential energy changes form and is released as heat energy.

A thermometer cannot measure potential energy. It can only measure energy due to motion of molecules. Temperature is a measure of the average kinetic energy of the molecules. Though common sense suggests that the temperature of a liquid should decrease continuously when it is cooled continuously, this does not happen at the freezing point.

Kinetic energy is the energy from the motion of molecules. There are three common temperature scales, Fahrenheit (F), Celsius (C), and Kelvin (K). The fixed points on the Fahrenheit and Celsius thermometers are based on the melting point of ice and the boiling point of water. It is assumed that ice melts at 32°F or 0°C and water boils at 212°F or 100°C at one atmospheric pressure. The more meaningful Kelvin scale is based on the lowest temperature possible to attain. The lowest possible temperature is –273°C. This temperature is called absolute zero or zero Kelvin. Scientists have succeeded in reaching very close to absolute zero but not absolute zero itself. According to this Kelvin scale (also called absolute scale), ice melts at 273 K and water boils at 373 K.

Sharp freezing points are obtained only for liquids that will give solid crystals. It is not possible to get sharp freezing points by cooling substances such as molten candle wax, molten butter and molten margarine.

Introduce students to the technique of drawing graphs and the advantages of graphs in interpreting and understanding results. Introduce the names, ordinate and abscissa, for the Y and X coordinates. Discuss the importance of labeling the axes and writing the units. They should learn to indicate data points as small circles or small squares instead of points, so that the data points can be easily recognized. They should learn the habit of using rulers to draw lines. It is extremely important for them to realize that a straight line graph need not touch all the data points. The points may be spread equally on both the sides of the line so that a best line is obtained. This is an important point and should be stressed often.

TEACHER'S SECTION
MAGIC WITH ICE CUBES

GOALS AND OBJECTIVES:

1. To find out whether you can lift an ice cube with a string without tying the ice cube to the string.

2. To find out whether you can cut through an ice cube without cutting it into two pieces.

3. To learn that ice melts when you add salt.

4. To learn that when pressure is applied to ice, it melts.

DISCUSSION BEFORE THE LABORATORY ACTIVITY:

Stress the need to wear an apron and safety goggles.

Discuss the properties of matter.

Discuss the phase changes, solid to liquid, liquid to gas, gas to liquid, and liquid to solid.

Discuss the reasons for putting salt on icy roads.

Adding a non volatile solute to a solvent lowers the freezing point of the solvent. This lowering of freezing point is a characteristic of the solvent. The lowering of freezing point depends on the number of solute particles in the solution and not on the nature of the solute particles. This is an example of a colligative property. That is, it is a property that depends on the number of solute particles and not on the nature of solute particles. One thousand grams of water containing 58.5 grams of common salt will freeze, not at 0°C, but at –3.72°C. In order for the 1000 grams of water to freeze at that temperature, you need 684 grams of sugar. Thus it is possible to put sugar on icy roads to melt the ice. However, it is much more expensive to use sugar instead of salt.

When pressure is applied on a block of ice, it melts. This helps the string to pass through the ice. When the pressure is relieved, ice is formed again.

TEACHER'S SECTION
BOILING POINT OF WATER

GOALS AND OBJECTIVES:

1. To find the nature of temperature change during phase change, liquid to gas.

2. To measure the boiling point of water.

3. To learn that the boiling of a liquid is a physical property.

4. To learn to represent data in a graph.

DISCUSSION BEFORE THE LABORATORY ACTIVITY:

It is important to stress the need to wear an apron and safety goggles.

Caution the students about handling the hot water with extreme care.

Discuss the properties of matter.

Discuss phase changes, solid to liquid, liquid to gas, gas to liquid, and liquid to solid. Introduce the fact that the boiling of a liquid and the condensation of its vapor can take place at the same temperature.

Temperature is a measure of the average kinetic energy of the molecules. Kinetic energy is the energy from the motion of molecules. There are three common temperature scales, Fahrenheit (F), Celsius (C), and Kelvin (K). The fixed points on the Fahrenheit and Celsius thermometers are based on the melting point of ice and the boiling point of water. It is assumed that ice melts at 32°F or 0°C and water boils at 212°F or 100°C at one atmospheric pressure. The more meaningful Kelvin scale is based on the lowest temperature possible to attain. The lowest possible temperature is –273°C. This temperature is called absolute zero or zero Kelvin. Scientists have succeeded in reaching very close to absolute zero but not absolute zero itself. According to this Kelvin scale (also called absolute scale), ice melts at 273 K and water boils at 373 K.

The normal boiling point of a liquid is the temperature at which the vapor pressure of the liquid is equal to normal atmospheric pressure. For example, the normal boiling point of water is 100°C. This means at 100°C, the vapor pressure of water is equal to one atmosphere. Water can be boiled at any temperature. The lower the atmospheric pressure, the lower the boiling point of water. Water boils at temperatures lower than 100°C on top of mountains because of lower atmospheric pressure conditions.

Ask students to name liquid substances that boil easily. Rubbing ethyl alcohol and rubbing isopropyl alcohol have lower boiling points than water.

When a liquid is heated uniformly, the temperature should increase uniformly. However, the temperature does not increase indefinitely. The temperature will increase uniformly until the boiling temperature of the liquid is reached. At the boiling temperature or boiling point, the temperature remains the same even if the substance is continuously heated. At the boiling point, the heat energy changes form and is stored as potential energy. (This is the reason why steam gives a more severe burn than boiling water. When the steam hits the skin, the stored potential energy, called, the heat of vaporization, is released as heat).

Whenever a liquid or a solution is to be boiled, always use some boiling chips or a glass stirrer in the cup or beaker. This helps to prevent the bumping of the liquid or solution.

A thermometer cannot measure potential energy. It can only measure the energy from the motion of molecules. Temperature is a measure of the average kinetic energy of the molecules. Though common sense, suggests that the temperature of a liquid should increase continuously on continuous heating, this does not happen at the boiling point.

Introduce students to the technique of drawing graphs and the advantages of graphs in interpreting and understanding results. Introduce the names, ordinate and abscissa, for the Y and X coordinates. Discuss the importance of labeling the axes and writing the units. They should learn to indicate data points as small circles or small squares instead of points, so that the data points can be easily recognized. They should learn the habit of using rulers for drawing lines. It is extremely important for them to realize that a straight line graph need not touch all the data points. The points may be spread equally on both the sides of the line so that a best line is obtained. This is an important point and should be stressed many times.

TEACHER'S SECTION
SPACES BETWEEN WATER MOLECULES IN LIQUID WATER

GOALS AND OBJECTIVES:

1. To find out whether there are empty spaces between water molecules in liquid water.

2. To find out whether these spaces can be occupied by other substances.

DISCUSSION BEFORE THE LABORATORY ACTIVITY:

It is very important to stress the need for wearing an apron and safety goggles.

Discuss some of the characteristics of water.

Discuss the solubility of solids such as salt and sugar in water.

Discuss the solubility of liquids such as wood alcohol (methanol), pure rubbing ethyl alcohol (ethyl alcohol or ethanol) and pure rubbing isopropyl alcohol (isopropyl alcohol or isopropanol or 2propanol) in water.

Discuss the solubility of kerosene, and gasoline in water.

It has been found experimentally that the total volume after mixing of two liquids can be greater than, equal to, or less than the sum of the individual volumes of the two liquids. The result depends on the nature of the two liquids and the attractive forces between the molecules of the two liquids.

It has been found that on mixing a solid with water, the total volume is quite often the volume of the water alone, provided the volume of water is very large compared to the volume of the solid. At fairly high concentrations, it has been found that the total volume is greater than the volume of the water, but less than the sum of the two volumes.

Water is a unique molecule. It has a bent structure, H – O , instead of a straight structure H – O – H.   H

The HOH bond angle is about 104 degrees. The water molecule is a covalent molecule. This means that the molecule is formed by the sharing of electrons. One electron from an oxygen atom and one electron from a hydrogen atom are shared by the two atoms in the molecule. Similarly another electron from the oxygen atom and an electron from the second hydrogen atom are shared by the two atoms in the molecule. However, the oxygen and hydrogen atoms have different abilities to attract electrons. The oxygen attracts the shared electrons much more strongly than the hydrogen atoms. This difference in attractive power for the electrons gives rise to partial electrical charges in the molecule. Oxygen has a partial negative charge and hydrogen has a partial positive charge. The special size, structure, and partial charges in the atoms of the molecule give rise to a special arrangement of molecules in liquid water and solid ice. Chemists call the attractive forces between the water molecules by a special name, hydrogen bond. It is really an attraction between the molecules due to partial charges in the molecule. The molecules are somewhat aligned in a tetrahedral way. These arrangements have subtle variations with changes in temperature. The special alignment of molecules in liquid water gives rise to empty spaces in the liquid water which can accommodate other solute molecules.

In this experiment the volume of the water is 20.0 mL. The volume after adding 2.0 grams of solid goes up to about 20.8 to 21.3 mL depending on the solid. After dissolving, the volume goes down by only about 0.1 to 0.3 mL. Therefore, the experiment has to be conducted carefully. Otherwise it is better to use about 10 grams of solid whose volume is measured using a 10.0 mL graduated cylinder and 200 mL water in a 250 mL graduated cylinder. It should be emphasized that the volume of the solid depends on the size of the crystals.

Normally the solubility of a salt is expressed as the maximum amount of salt that can be dissolved in 100 grams of water at a specified temperature. The solubility of baking soda (sodium hydrogen carbonate, $NaHCO_3$) is 6.9 grams at $0°C$ and 16.4 grams at $60°C$ in 100 mL water. Other salts and sugar have comparatively much higher solubilities. Therefore, there may be some difficulty in dissolving the 2.0 grams of baking soda in 20.0 mL water. If the temperature is about $25°C$, there should be no problem. Otherwise, a salt such as sodium acetate trihydrate (salt of vinegar) may be used instead of baking soda.

TEACHER'S SECTION
MAGIC WITH "THE HEAT SOLUTION"

GOALS AND OBJECTIVES:

1. To determine the amount of heat produced by rapid crystallization of a substance.

2. To learn to use a mathematical formula for computing the number of calories of heat produced.

3. To learn that the process of freezing releases energy.

DISCUSSION BEFORE THE LABORATORY ACTIVITY:

It is very important to stress the need for wearing an apron and safety goggles.

Discuss heat energy and the flow of heat energy.

Discuss the meaning of temperature.

Discuss the definitions of calorie and specific heat.

Discuss the formula for the number of calories.

Discuss exothermic and endothermic processes.

Temperature is a measure of the average kinetic energy of the molecules. Kinetic energy is the energy from the motion of molecules. There are three common temperature scales, Fahrenheit (F), Celsius (C), and Kelvin (K). The fixed points on the Fahrenheit and Celsius thermometers are based on the melting point of ice and the boiling point of water. It is assumed that ice melts at 32°F or 0°C and water boils at 212°F or 100°C at one atmospheric pressure. The more meaningful Kelvin scale is based on the lowest temperature possible to attain. The lowest possible temperature is –273°C. This temperature is called absolute zero or zero Kelvin. Scientists have succeeded in reaching very close to absolute zero but not absolute zero itself. According to this Kelvin scale (also called absolute scale), ice melts at 273 K and water boils at 373 K.

One calorie is the amount of heat needed to raise the temperature of 1 gram of water by 1° (Celsius). The exact definition of calorie specifies the temperature change of the water from 14.5 to 15.5 °C.

This calorie is different from the food calorie. One food calorie = 1000 calories = 1 kilocalorie or 1 kcal.

Nowadays scientists use another unit, joule, for energy. One joule = 4.184 calories. One joule is the energy which is equal to the work done by a force of one Newton acting over a distance of one meter.

Specific heat is the heat required to raise the temperature of 1 gram of any substance by 1°C.

The heat of fusion of ice is 80 calories per gram. This means that when one gram of water at 0°C freezes into ice, it releases 80 calories of heat. This is not a small amount. Most substances have much less heat of fusion. Even though ice seems to melt easily, it is one of the most difficult substances to melt. One gram of iron can be melted more easily at its melting point with less heat.

In an exothermic reaction or process, energy is released as heat. This energy comes from the stored energy in the substance.

In an endothermic reaction or process, energy is absorbed and stored as potential energy.

The formula for sodium acetate trihydrate is $CH_3COONa \cdot 3H_2O$. This indicates an association of 3 molecules of water with one molecule of sodium acetate in the crystal. In some books, this is also written as $NaC_2H_3O_2 \cdot 3H_2O$.

TEACHER'S SECTION
MAGIC WITH A COLD PACK

GOALS AND OBJECTIVES:

1. To find out whether the cooling produced by dissolving a substance in water is sufficient to freeze some pure water.

2. To learn that dissolving a substance in water can make the water cold.

3. To learn that the cooling produced by dissolving a solid in water can be used to freeze pure water.

DISCUSSION BEFORE THE LABORATORY ACTIVITY:

It is very important to stress the need for wearing an apron and safety goggles.

Discuss heat energy and the flow of heat energy.

Discuss conductors and insulators of heat. Ask the students to name some insulators and conductors of heat.

Discuss the meaning of temperature.

Discuss exothermic and endothermic processes.

Temperature is a measure of the average kinetic energy of the molecules. Kinetic energy is the energy from the motion of molecules. There are three common temperature scales, Fahrenheit (F), Celsius (C), and Kelvin (K). The fixed points on the Fahrenheit and Celsius thermometers are based on the melting point of ice and the boiling point of water. It is assumed that ice melts at 32°F or 0°C and water boils at 212°F or 100°C at one atmospheric pressure. The more meaningful Kelvin scale is based on the lowest temperature possible to attain. The lowest possible temperature is –273°C. This temperature is called absolute zero or zero Kelvin. Scientists have succeeded in reaching very close to absolute zero but not absolute zero itself. According to this Kelvin scale (also called absolute scale), ice melts at 273 K and water boils at 373 K.

One calorie is the amount of heat needed to raise the temperature of 1 gram of water by 1° (Celsius). The exact definition of calorie specifies the temperature change of the water from 14.5 to 15.5 °C.

This calorie is different from the food calorie. One food calorie = 1000 calories = 1 kilocalorie or 1 kcal.

Nowadays scientists use another unit, joule, for energy. One joule = 4.184 calories. One joule is the energy which is equal to the work done by a force of one Newton acting over a distance of one meter.

Specific heat is the heat required to raise the temperature of 1 gram of any substance by 1°C.

The heat of fusion of ice is 80 calories per gram. This means that when one gram of water at 0°C freezes into ice, it releases 80 calories of heat. This is not a small amount. Most substances have much less heat of fusion. Even though ice seems to melt easily, it is one of the most difficult substances to melt. One gram of iron can be melted more easily at its melting point with less heat.

In an exothermic reaction or process, energy is released as heat. This energy comes from the stored energy in the substance.

In an endothermic reaction or process, energy is absorbed and stored as potential energy.

The formula for ammonium nitrate is NH_4NO_3. The cold pack contains solid ammonium nitrate in an outer bag and liquid water in an inside bag. Squeezing the inside bag will allow the water to mix with the solid. When the solid dissolves in the water, it absorbs heat from the surroundings and changes into potential energy. This results in a lowering of temperature. Thus the dissolution of ammonium nitrate in water is an endothermic process. Every gram of ammonium nitrate absorbs about 78 calories of heat during the dissolving process. This means that one gram of ammonium nitrate can lower the temperature of 1 gram of water by about 78°C. Commercial cold packs contain about 200 grams of ammonium nitrate and about 200 mL water. The solubility of ammonium nitrate is about 50 grams in 100 mL water at 0°C. This means, it is not possible to dissolve all the ammonium nitrate in the amount of water in the cold pack. **Ammonium nitrate is a powerful oxidizing agent. Contamination with combustible materials can result in explosion. It is also toxic if ingested. Use proper disposal procedures recommended by the school.**

It is not possible to freeze the water between the wooden block and the beaker unless the starting temperature of the water is about 10°C or lower. It is preferable not to do the trick experiment during summer. It works better on cold days.

The purpose of the wooden block is to act as an insulator so that the water–ammonium nitrate system can withdraw heat from the water on the wooden block. If a metal block is used instead of the wooden block, it conducts away the heat. In such a case, it may not be possible to freeze the water on the block unless a small metal block with low heat capacity is used.

TEACHER'S SECTION
MAGIC WITH SUPER-ABSORBENT DIAPERS

GOALS AND OBJECTIVES:

1. To determine the amount of water that can be absorbed by a known amount of the super-absorbent material used in disposable diapers.

2. To learn that some substances swell in water and can hold a substantial quantity of water in the form of a gel.

DISCUSSION BEFORE THE LABORATORY ACTIVITY:

It is very important to stress the need for wearing an apron and safety goggles.

Discuss the meaning of hygroscopic materials.

Discuss the meaning of deliquescent materials.

Discuss the meaning of water of hydration in a crystal.

Discuss the formation of gels.

Credit for the development of super-absorbent materials goes to the United States Department of Agriculture. These are used as agricultural thickeners and as additives for secondary oil recovery operations. A super-absorbent, waterlock (sodium polyacrylate), is available from Flinn Scientific Inc. A gram of waterlock can absorb about 80 grams of water.

150 Fundamentals of Science Activity Series

TEACHER'S SECTION
ABSORPTION OF MOISTURE FROM AIR

GOALS AND OBJECTIVES:

1. To determine whether chemicals absorb moisture from the air.
2. To learn that there is moisture in air.
3. To learn that chemicals that absorb moisture can be used to reduce humidity in the air.

DISCUSSION BEFORE THE LABORATORY ACTIVITY:

It is very important to stress the need for wearing an apron and safety goggles.

Discuss the meaning of hygroscopic materials.

Discuss the meaning of deliquescent materials.

Discuss the meaning of water of hydration in a crystal.

Discuss color changes associated with water of hydration.

Discuss the formation of gels.

Discuss ways of controlling humidity.

Discuss ways of protecting materials from dampness.

Discuss ways of controlling dust on dirt roads.

It is preferable to do this experiment on days of high humidity.

Calcium chloride and cobalt chloride used for this experiment must be anhydrous. If anhydrous materials are not available, they can be produced by heating small amounts in petri dishes and covering them quickly before cooling.

Anhydrous cobalt chloride is blue. It forms a hydrate with 6 water molecules. The hydrated cobalt chloride, $CoCl_2 \cdot 6H_2O$, is red. Anhdrous copper sulfate is white or light green. It forms a hydrate with 5 water molecules. The hydrated copper sulfate, $CuSO_4 \cdot 5H_2O$ is blue.

Cobalt chloride and copper sulphate are toxic if ingested. Collect waste in a bottle and use disposal procedures recommended by the school.

TEACHER'S SECTION
EXPANSION OF AN EGG

GOALS AND OBJECTIVES:

1. To determine the increase in volume of an egg because of osmosis.
2. To learn that the shell of an egg reacts with vinegar.
3. To learn that water enters inside an egg with no shell.
4. To learn that the size of an egg with no shell increases when it is kept in water.

DISCUSSION BEFORE THE LABORATORY ACTIVITY:

It is very important to stress the need to wear an apron and safety goggles.

Caution students about handling raw eggs.

Discuss chemical reactions.

Discuss the composition of the shell of an egg.

Discuss ways of removing the shell from a raw egg.

Discuss osmosis.

Discuss how plants and trees pick up water from the ground.

The shell of an egg is mostly calcium carbonate. All carbonates decompose by reacting with acids. Vinegar is a solution of 5% acetic acid, CH_3COOH. All acids produce hydrogen ions, $H^+(aq)$ or hydronium ions, H_3O^+, in water. Vinegar reacts with the shell and produces soluble calcium acetate, water and carbon dioxide. The reaction is:

$$CaCO_3 + 2\ CH_3COOH \longrightarrow Ca(OOCCH_3)_2 + H_2O + CO_2$$

Once the shell is removed, water passes through the membrane in the egg and enters inside the egg because of a process called osmosis. This increases the size of the egg.

152 Fundamentals of Science Activity Series

TEACHER'S SECTION
DECOLORIZATION WITH ACTIVATED CHARCOAL

GOALS AND OBJECTIVES:

1. To remove food colors from water using activated charcoal.
2. To learn the properties of activated charcoal.
3. To learn a technique for purifying water.
4. To learn the technique of filtration.

DISCUSSION BEFORE THE LABORATORY ACTIVITY:

It is very important to stress the need for wearing an apron and safety goggles.

Discuss color and the color of the food dyes.

Discuss the solubility of different substances in water.

Discuss water pollution.

Discuss ways to purify water.

Discuss the properties of charcoal.

Discuss the various uses of charcoal.

Teach the technique of filtration.

TEACHER'S SECTION
MILK AND FOOD COLORS

GOALS AND OBJECTIVES:

1. To determine which food color mixes more easily with milk.
2. To learn the influence of detergent on the mixing process.

DISCUSSION BEFORE THE LABORATORY ACTIVITY:

It is very important to stress the need for wearing an apron and safety goggles.

Discuss color and the color of the food dyes.

Find the F D & C No. of the food colors from the packages.

Discuss the composition of milk.

Discuss the properties of soap and detergents.

Discuss the solubility of different substances in water and in milk.

Ask students to produce other colors using red, blue and yellow dyes. Green is a mixture of yellow and blue dyes. Demonstrate this by mixing one drop each of blue and yellow food colors in a clear plastic cup. Add about 50 mL water to this and observe.

Show the complicated structures of the dyes to the students. These structures are not easily available from text books. The chemical structures of the most commonly used food dyes are given at the beginning of the book, under the title, "Chemical Structures".

The mixing of different dyes with water or milk depends on the chemical structure of the dyes. It is very hard to judge the differences in the behavior of their mixing with water or milk by looking at their complicated chemical structures. It is much easier to do the experiment and find out the answers. The mixing of solid dyes with water or other solvents such as alcohol provides further evidence for their different characteristics.

The unique structure of water is responsible for its high surface tension. Adding a small amount of detergent lowers the surface tension considerably and helps the mixing process.

The solubility of a solid or a liquid in water depends on several factors. It depends on the nature of water, the nature of chemical bonding in the solid or liquid, and their mutual compatibilities. Only similar substances dissolve in similar solvents. For example sugar and water are similar. A water molecule has partial electrical charges in its atoms. The sugar molecule also has partial electrical charges in its atoms. The partial positive charge on the hydrogen atom of the water molecule can attract the partial negative charge on the oxygen atom in the sugar molecule. The partial negative charge on the oxygen atom in the water molecule can attract the partial positive charge on the hydrogen atom in the sugar molecule. These attractive forces dominate and will result in the dissolution of sugar in water.

Similar reasoning explains the dissolution of salt and water. However, salt is an ionic solid, with a +1 electrical charge on the sodium atom and a −1 electrical charge on the chlorine atom instead of partial electrical charges.

A substance such as gasoline has practically no partial electrical charges and therefore, is not attracted to the partial electrical charges on the atoms of the water molecule. Therefore, gasoline and water do not mix. Substances such as ethanol or ethyl alcohol have one part of the molecule which is like gasoline and another part of the molecule which is like water. The attraction of the latter part with water dominates and therefore, alcohol is soluble in water. Substances such as baby oil and cooking oil also have two parts. However, the major portions of their molecules are similar to gasoline and therefore, these oils do not mix with water. The iodine used in Activity 38, *Molecules of Solids that Do Not Mix*, is similar to gasoline in that it has no partial electrical charges in the molecule. Therefore, it does not dissolve in water. Instead it dissolves in a similar substance such as baby oil. The solid food colors have several atoms with partial electrical charges and their attractions with water dominate so that they are soluble in water.

The food colors are not true molecules. To be a true molecule, the substance must have covalent bonding only (sharing of electrons) between their atoms. The commonly used food colors are ionic substances because they have positive sodium ions and negative ions at some ends of the huge molecule. However, the food colors have been extensively used by many investigators as true molecules. This is partly justified because the charges are negligible compared to the huge size of the molecules.

TEACHER'S SECTION
MIXING MOLECULES AND WATER IN DIFFERENT CONTAINERS

GOALS AND OBJECTIVES:

1. To determine the ease of mixing of molecules in different containers.
2. To select the best container that gives maximum information when studying the mixing of molecules.
3. To learn to identify different glassware.

DISCUSSION BEFORE THE LABORATORY ACTIVITY:

It is very important to stress the need for wearing an apron and safety goggles.

Practice identification of different glassware such as beaker, test tube, graduated cylinder, petri dish, Erlenmeyer flask, and filtering flask with side tube.

Discuss color and the color of the food dyes.

Find the F D & C No. of the food colors from the packages.

Discuss the solubility of different substances in water.

Salt dissolves in water. But oil does not dissolve in water. Why?

The solubility of a solid or a liquid in water depends on several factors. It depends on the nature of water, the nature of chemical bonding in the solid or liquid, and their mutual compatibilities. Only similar substances dissolve in similar solvents. For example sugar and water are similar. A water molecule has partial electrical charges in its atoms. The sugar molecule also has partial electrical charges in its atoms. The partial positive charge on the hydrogen atom of the water molecule can attract the partial negative charge on the oxygen atom in the sugar molecule. The partial negative charge on the oxygen atom in the water molecule can attract the partial positive charge on the hydrogen atom in the sugar molecule. These attractive forces dominate and will result in the dissolution of sugar in water.

Similar reasoning explains the dissolution of salt and water. However, salt is an ionic solid, with a +1 electrical charge on the sodium atom and a −1 electrical charge on the chlorine atom instead of partial electrical charges.

A substance such as gasoline has practically no partial electrical charges and therefore, is not attracted to the partial electrical charges on the atoms of the water molecule. Therefore, gasoline and water do not mix. Substances such as ethanol or ethyl alcohol have one part of the molecule which is like gasoline and another part of the molecule which is like water. The attraction of the latter part with water dominates and therefore, alcohol is soluble in water. Substances such as baby oil and cooking oil also have two parts. However, the major portions of their molecules are similar to gasoline and therefore, these oils do not mix with water. The iodine used in Activity 38, *Molecules of Solids that Do Not Mix*, is similar to gasoline in that it has no partial electrical charges in the molecule. Therefore, it does not dissolve in water. Instead it dissolves in a similar substance such as baby oil. The solid food colors have several atoms with partial electrical charges and their attractions with water dominate so that they are soluble in water.

The food colors are not true molecules. To be a true molecule, the substance must have covalent bonding only (sharing of electrons) between their atoms. The commonly used food colors are ionic substances because they have positive sodium ions and negative ions at some ends of the huge molecule. However, the food colors have been extensively used by many investigators as true molecules. This is partly justified because the charges are negligible compared to the huge size of the molecules.

Discuss diffusion.

Show the complicated structures of the dyes to the students. These structures are not easily available from text books. The chemical structures of the most commonly used food dyes are given at the beginning of the book, under the title "Chemical Structures".

The mixing of different dyes with water depends on the chemical structures of the dyes. It is hard to judge the differences in the behavior of their mixing with water by looking at their complicated chemical structures. It is much easier to do the experiment and find out the answers. The mixing of solid dyes with water or other solvents such as alcohol provides further evidence for their different characteristics.

Sometimes the process of mixing can be more easily observed in shallow dishes. Sometimes you need a container with depth.

TEACHER'S SECTION
MIXING FOOD COLOR AND WATER

GOALS AND OBJECTIVES:

1. To determine which food color mixes most easily with water.
2. To determine the time needed for the mixing of food colors and water.

DISCUSSION BEFORE THE LABORATORY ACTIVITY:

It is very important to stress the need for wearing an apron and safety goggles.

Discuss color and the color of the food dyes.

Find the F D & C No. of the food colors from the packages.

Discuss the solubility of different substances in water.

The solubility of a solid or a liquid in water depends on several factors. It depends on the nature of water, the nature of chemical bonding in the solid or liquid, and their mutual compatibilities. Only similar substances dissolve in similar solvents. For example sugar and water are similar. A water molecule has partial electrical charges in its atoms. The sugar molecule also has partial electrical charges in its atoms. The partial positive charge on the hydrogen atom of the water molecule can attract the partial negative charge on the oxygen atom in the sugar molecule. The partial negative charge on the oxygen atom in the water molecule can attract the partial positive charge on the hydrogen atom in the sugar molecule. These attractive forces dominate and will result in the dissolution of sugar in water.

Similar reasoning explains the dissolution of salt and water. However, salt is an ionic solid, with a +1 electrical charge on the sodium atom and a −1 electrical charge on the chlorine atom instead of partial electrical charges.

A substance such as gasoline has practically no partial electrical charges and therefore, is not attracted to the partial electrical charges on the atoms of the water molecule. Therefore, gasoline and water do not mix. Substances such as ethanol or ethyl alcohol have one part of the molecule which is like gasoline and another part of the molecule which is like water. The attraction of the latter part with water dominates and therefore, alcohol is soluble in water. Substances such as baby oil and cooking oil also have two parts. However, the major portions of their molecules are similar to gasoline and therefore, these oils do not mix with water. The iodine used in Activity 38, *Molecules of Solids that Do Not Mix*, is similar to gasoline in that it has no partial electrical charges in the molecule. Therefore, it does not dissolve in water. Instead it dissolves in a similar substance such as baby oil. The solid food colors have several atoms with partial electrical charges and their attractions with water dominate so that they are soluble in water.

The food colors are not true molecules. To be a true molecule, the substance must have covalent bonding only (sharing of electrons) between their atoms. The commonly used food colors are ionic substances because they have positive sodium ions and negative ions at some ends of the huge molecule. However, the food colors have been extensively used by many investigators as true molecules. This is partly justified because the charges are negligible compared to the huge size of the molecules.

Discuss diffusion. The spreading of a substance because of molecular motion is called diffusion. Diffusion is faster at higher temperatures.

The motion of molecules in a solid is very slow. The motion of molecules in a gas is very rapid. The motion of molecules in a liquid is between that of a solid and a gas.

Show the complicated structures of the dyes to the students. These structures are not easily available from text books. The chemical structures of the most commonly used food dyes are given at the beginning of the book, under the title "Chemical Structures".

The mixing of different dyes with water depends on the chemical structures of the dyes. It is very hard to judge the differences in the behavior of their mixing with water by looking at their complicated chemical structures. It is much easier to do the experiment and find out the answers. The mixing of solid dyes with water or other solvents such as alcohol provides further evidence of their different characteristics.

The complete spreading of colors takes a lot of time (two to three hours). There is no need to wait for this to happen. The trends in the spreading of different colors can be seen in less than 5 minutes. The experiment is much faster, if done in 50 mL graduated cylinders. The relative spreading of the colors can be easily seen in two to three minutes and most of the spreading of the colors takes place in 5 to 10 minutes. Also the complete spreading is much faster in the graduated cylinders than in the clear plastic cups. When a clear plastic cup is used, more of the concentrated dye lies at the bottom of the cup. When a graduated cylinder is used, there is a downward spreading because of the higher density of the food color solution. The order of spreading seems to be yellow >red >green >blue.

TEACHER'S SECTION
MIXING A SOLID AND A LIQUID

GOALS AND OBJECTIVES:

1. To find out how easy or difficult it is for molecules of solid food colors to spread evenly in water.

2. To learn that molecular motion in solids is slower than in liquids.

3. To determine the time needed for the mixing of solid food color and water.

DISCUSSION BEFORE THE LABORATORY ACTIVITY:

It is very important to stress the need to wear an apron and safety goggles.

Discuss color and the color of the food dyes.

Find the F D & C No. of the food colors from the packages.

Discuss the solubility of different substances in water. The solubility of a solid or a liquid in water depends on several factors. It depends on the nature of water, the nature of chemical bonding in the solid or liquid, and their mutual compatibilities. Only similar substances dissolve in similar solvents. For example sugar and water are similar. A water molecule has partial electrical charges in its atoms. The sugar molecule also has partial electrical charges in its atoms. The partial positive charge on the hydrogen atom of the water molecule can attract the partial negative charge on the oxygen atom in the sugar molecule. The partial negative charge on the oxygen atom in the water molecule can attract the partial positive charge on the hydrogen atom in the sugar molecule. These attractive forces dominate and will result in the dissolution of sugar in water.

Similar reasoning explains the dissolution of salt and water. However, salt is an ionic solid, with a +1 electrical charge on the sodium atom and a −1 electrical charge on the chlorine atom instead of partial electrical charges.

A substance such as gasoline has practically no partial electrical charges and therefore, is not attracted to the partial electrical charges on the atoms of the water molecule. Therefore, gasoline and water do not mix. Substances such as ethanol or ethyl alcohol have one part of the molecule which is like gasoline and another part of the molecule which is like water. The attraction of the latter part with water dominates and therefore, alcohol is soluble in water. Substances such as baby oil and cooking oil also have two parts. However, the major portions of their molecules are similar to gasoline and therefore, these oils do not mix with water. The iodine used in Activity 38, *Molecules of Solids that Do Not Mix*, is similar to gasoline in that it has no partial electrical charges in the molecule. Therefore, it does not dissolve in water. Instead it dissolves in a similar substance such as baby oil. The solid food colors have several atoms with partial electrical charges and their attractions with water dominate so that they are soluble in water.

The food colors are not true molecules. To be a true molecule, the substance must have covalent bonding only (sharing of electrons) between their atoms. The commonly used food colors are ionic substances because they have positive sodium ions and negative ions at some ends of the huge molecule. However, the food colors have been extensively used by many investigators as true molecules. This is partly justified because the charges are negligible compared to the huge size of the molecules.

Discuss diffusion.

The motion of molecules in a solid is very slow. The motion of molecules in a gas is very fast. The motion of molecules in a liquid is between that of a solid and a gas.

The spreading of a substance because of molecular motion is called diffusion. Diffusion is faster at higher temperatures.

The solid food colors may be obtained by powdering the food color tablets available during Easter time. For a real comparison with liquid food colors, it is better to dry some liquid food colors in petri dishes (keep them in open air for a few days) and scrape the dried colors to get the powders. Store these powders in small vials.

The mixing of different dyes with water depends on the chemical structures of the dyes. It is very hard to judge the differences in their mixing with water by looking at their complicated chemical structures. It is much easier to do the experiment and find out the answers. The mixing of solid dyes with water or other solvents such as alcohol provides further evidence for their different characteristics.

The red and blue solid food colors seem to spread initially on the surface of water and then move down in streaks. This streaking is quite spectacular to watch. The solid yellow food color does not seem to do this and spreads rather fast. The spreading on the surface of water is greatest for the solid blue food color.

This experiment is very exciting to watch if done in petri dishes. The order of spreading is yellow >red >blue.

TEACHER'S SECTION
MOVEMENT OF MOLECULES IN SOLIDS

GOALS AND OBJECTIVES:

1. To determine the ease with which molecules mix when added to ice.
2. To learn that molecular motion in solids is much slower than in liquids.
3. To determine the time needed for the mixing of solid food color with ice.

DISCUSSION BEFORE THE LABORATORY ACTIVITY:

It is very important to stress the need to wear an apron and safety goggles.

Discuss color and the color of the food dyes.

Find the F D & C No. of the food colors from the packages.

Discuss the solubility of different substances in water.

The solubility of a solid or a liquid in water depends on several factors. It depends on the nature of water, the nature of chemical bonding in the solid or liquid, and their mutual compatibilities. Only similar substances dissolve in similar solvents. For example sugar and water are similar. A water molecule has partial electrical charges in its atoms. The sugar molecule also has partial electrical charges in its atoms. The partial positive charge on the hydrogen atom of the water molecule can attract the partial negative charge on the oxygen atom in the sugar molecule. The partial negative charge on the oxygen atom in the water molecule can attract the partial positive charge on the hydrogen atom in the sugar molecule. These attractive forces dominate and will result in the dissolution of sugar in water.

Similar reasoning explains the dissolution of salt and water. However, salt is an ionic solid, with a +1 electrical charge on the sodium atom and a –1 electrical charge on the chlorine atom instead of partial electrical charges.

A substance such as gasoline has practically no partial electrical charges and therefore, is not attracted to the partial electrical charges on the atoms of the water molecule. Therefore, gasoline and water do not mix. Substances such as ethanol or ethyl alcohol have one part of the molecule which is like gasoline and another part of the molecule which is like water. The attraction of the latter part with water dominates and therefore, alcohol is soluble in water. Substances such as baby oil and cooking oil also have two parts. However, the major portions of their molecules are similar to gasoline and therefore, these oils do not mix with water. The iodine used in Activity 38, *Molecules of Solids that Do Not Mix*, is similar to gasoline in that it has no partial electrical charges in the molecule. Therefore, it does not dissolve in water. Instead it dissolves in a similar substance such as baby oil. The solid food colors have several atoms with partial electrical charges and their attractions with water dominate so that they are soluble in water.

The food colors are not true molecules. To be a true molecule, the substance must have covalent bonding only (sharing of electrons) between their atoms. The commonly used food colors are ionic substances because they have positive sodium ions and negative ions at some ends of the huge molecule. However, the food colors have been extensively used by many investigators as true molecules. This is partly justified because the charges are negligible compared to the huge size of the molecules.

Discuss the possibility of mixing different substances in ice.

Ask the reasons for the difficulty in mixing two solids compared to two liquids.

Discuss molecular motion in solids, liquids, and gases and ways of determining their relative motions.

The motion of molecules in a solid is very slow. The motion of molecules in a gas is very fast. The motion of molecules in a liquid is between that of a solid and a gas.

The spreading of a substance because of molecular motion is called diffusion. Diffusion is faster at higher temperatures.

The solid food colors may be obtained by powdering the food color tablets available during Easter time. For a real comparison with liquid food colors, it is better to dry some liquid food colors in petri dishes (keep them in open air for a few days) and scrape the dried colors to get the powders. Store these powders in small vials.

The mixing of different dyes with ice depends on the chemical structures of the dyes. It is very hard to judge the differences in the behavior of their mixing with ice by looking at their complicated chemical structures. It is much easier to do the experiment and find out the answers. It will be difficult to get any noticeable mixing between the solid food color and ice. Mixing will be noticeable only when the ice starts melting.

It is interesting to watch the streaks of blue and red food colors through the ice. This does not happen for the yellow food color. In order for the experiment to work successfully, it is better to have large ice cubes or blocks of ice. Water frozen in plastic petri dishes is a very good substitute. With small ice cubes, the liquid food colors seem to flow down easily into the petri dishes.

TEACHER'S SECTION
MIXING MOLECULES IN A SOLID AND WATER AT DIFFERENT TEMPERATURES

GOALS AND OBJECTIVES:

1. To study the motion of molecules in solids on mixing with water at different temperatures.

2. To learn that molecules move much faster at higher temperatures than at lower temperatures.

DISCUSSION BEFORE THE LABORATORY ACTIVITY:

It is very important to stress the need for wearing an apron and safety goggles.

Discuss color and the color of the food dyes.

Find the F D & C No. of the food colors from the packages.

Discuss the solubility of different substances in water. The solubility of a solid or a liquid in water depends on several factors. It depends on the nature of water, the nature of chemical bonding in the solid or liquid, and their mutual compatibilities. Only similar substances dissolve in similar solvents. For example sugar and water are similar. A water molecule has partial electrical charges in its atoms. The sugar molecule also has partial electrical charges in its atoms. The partial positive charge on the hydrogen atom of the water molecule can attract the partial negative charge on the oxygen atom in the sugar molecule. The partial negative charge on the oxygen atom in the water molecule can attract the partial positive charge on the hydrogen atom in the sugar molecule. These attractive forces dominate and will result in the dissolution of sugar in water.

Similar reasoning explains the dissolution of salt and water. However, salt is an ionic solid, with a +1 electrical charge on the sodium atom and a −1 electrical charge on the chlorine atom instead of partial electrical charges.

A substance such as gasoline has practically no partial electrical charges and therefore, is not attracted to the partial electrical charges on the atoms of the water molecule. Therefore, gasoline and water do not mix. Substances such as ethanol or ethyl alcohol have one part of the molecule which is like gasoline and another part of the molecule which is like water.

The attraction of the latter part with water dominates and therefore, alcohol is soluble in water. Substances such as baby oil and cooking oil also have two parts. However, the major portions of their molecules are similar to gasoline and therefore, these oils do not mix with water. The iodine used in Activity 38, *Molecules of Solids that Do Not Mix*, is similar to gasoline in that it has no partial electrical charges in the molecule. Therefore, it does not dissolve in water. Instead it dissolves in a similar substance such as baby oil. The solid food colors have several atoms with partial electrical charges and their attractions with water dominate so that they are soluble in water.

The food colors are not true molecules. To be a true molecule, the substance must have covalent bonding only (sharing of electrons) between their atoms. The commonly used food colors are ionic substances because they have positive sodium ions and negative ions at some ends of the huge molecule. However, the food colors have been extensively used by many investigators as true molecules. This is partly justified because the charges are negligible compared to the huge size of the molecules.

Discuss the possible relationship between solubility and temperature.

Ask the reasons for the difficulty in mixing two solids compared to two liquids.

Discuss molecular motion in solids, liquids, and gases and ways of determining their relative motions.. The motion of molecules in a solid is very slow. The motion of molecules in a gas is very fast. The motion of molecules in a liquid is inbetween that of a solid and a gas. The spreading of a substance due to molecular motion is called diffusion. Diffusion is faster at higher temperatures.

The solid food colors may be obtained by powdering the food color tablets available during Easter time. For a real comparison with liquid food colors, it is better to dry some liquid food colors in petri dishes (keep them in open air for a few days) and scrape the dried colors to get the powders. Store these powders in small vials.

The mixing of different dyes with water depends on the chemical structures of the dyes. It is very hard to judge the differences in the behavior of their mixing with ice by looking at their complicated chemical structures. It is much easier to do the experiment and find out the answers.

The red and blue solid food colors seem to spread initially on the surface of water and then move down in streaks. This streaking is quite spectacular to watch. The solid yellow food color does not seem to do this and spreads rather fast. The spreading on the surface of water is greatest for the solid blue food color. This experiment is very exciting to watch if done in petri dishes. The order of spreading is yellow >red >blue.

TEACHER'S SECTION
MIXING MOLECULES IN A SOLUTION AND WATER AT DIFFERENT TEMPERATURES

GOALS AND OBJECTIVES:

1. To study the motion of molecules in a solution as they mix with water at different temperatures.

2. To learn that molecules move much faster at higher temperatures than at lower temperatures.

DISCUSSION BEFORE THE LABORATORY ACTIVITY:

It is very important to stress the need to wear an apron and safety goggles.

Discuss color and the color of the food dyes.

Find the F D & C No. of the food colors from the packages.

Discuss the solubility of different substances in water.

The solubility of a solid or a liquid in water depends on several factors. It depends on the nature of water, the nature of chemical bonding in the solid or liquid, and their mutual compatibilities. Only similar substances dissolve in similar solvents. For example sugar and water are similar. A water molecule has partial electrical charges in its atoms. The sugar molecule also has partial electrical charges in its atoms. The partial positive charge on the hydrogen atom of the water molecule can attract the partial negative charge on the oxygen atom in the sugar molecule. The partial negative charge on the oxygen atom in the water molecule can attract the partial positive charge on the hydrogen atom in the sugar molecule. These attractive forces dominate and will result in the dissolution of sugar in water.

Similar reasoning explains the dissolution of salt and water. However, salt is an ionic solid, with a +1 electrical charge on the sodium atom and a −1 electrical charge on the chlorine atom instead of partial electrical charges.

A substance such as gasoline has practically no partial electrical charges and therefore, is not attracted to the partial electrical charges on the atoms of the water molecule. Therefore, gasoline and water do not mix. Substances such as ethanol or ethyl alcohol have one part of the molecule which is like gasoline and another part of the molecule which is like water. The attraction of the latter part with water dominates and therefore, alcohol is soluble in water. Substances such as baby oil and cooking oil also have two parts. However, the major portions of their molecules are similar to gasoline and therefore, these oils do not mix with water. The iodine used in Activity 38, *Molecules of Solids that Do Not Mix,* is similar to gasoline in that it has no partial electrical charges in the molecule. Therefore, it does not dissolve in water. Instead it dissolves in a similar substance such as baby oil. The solid food colors have several atoms with partial electrical charges and their attractions with water dominate so that they are soluble in water.

The food colors are not true molecules. To be a true molecule, the substance must have covalent bonding only (sharing of electrons) between their atoms. The commonly used food colors are ionic substances because they have positive sodium ions and negative ions at some ends of the huge molecule. However, the food colors have been extensively used by many investigators as true molecules. This is partly justified because the charges are negligible compared to the huge size of the molecules.

Discuss the possible relationship between solubility and temperature.

Ask for the reasons for the difficulty in mixing two solids compared to two liquids.

Discuss molecular motion in solids, liquids, and gases and ways of determining their relative motions..

The motion of molecules in a solid is very slow. The motion of molecules in a gas is very fast. The motion of molecules in a liquid is intermediate between that of a solid and a gas.

The spreading of a substance due to molecular motion is called diffusion. Diffusion is faster at higher temperatures.

Solid food colors may be obtained by powdering the food color tablets available during Easter time. For a real comparison with liquid food colors, it is better to dry some liquid food colors in petri dishes (keep them in open air for a few days) and scrape the dried colors to get the powders. Store these powders in small vials.

The way different dyes mix with water depends on the chemical structures of the dyes. It is very hard to judge the differences in the ways they mix with ice by looking at their complicated chemical structures. It is much easier to do the experiment and find out the answers.

The complete uniform spreading of colors takes a lot of time (two to three hours). There is no need to wait for this to happen. The trends in the spreading of different colors can be seen in less than 5 minutes. The experiment is much faster, if done in 100 mL graduated cylinders. The relative spreading of the colors can be easily seen in two to three minutes and most of the spreading of the colors takes place in 5 to 10 minutes. Also the complete spreading is much faster in the graduated cylinders than in the clear plastic cups. When a clear plastic cup is used, more of the concentrated dye lies at the bottom of the cup. When a graduated cylinder is used, there is a downward spreading because of the higher density of the food color solution. The order of spreading seems to be yellow >red >green >blue.

TEACHER'S SECTION
MOLECULES OF SOLIDS THAT DO NOT MIX

GOALS AND OBJECTIVES:

1. To determine whether molecules of solids dissolve in all liquids.

2. To determine whether molecules of solids mix with the molecules of all liquids.

3. To learn that some substances that do not dissolve in water will dissolve in other liquids.

4. To learn that some substances that dissolve in water do not dissolve in other liquids.

DISCUSSION BEFORE THE LABORATORY ACTIVITY:

It is very important to stress the need for wearing an apron and safety goggles.

Discuss the solubility of solids such as salt and sugar in water.

Discuss the solubility of liquids such as rubbing ethyl alcohol and rubbing isopropyl alcohol in water.

Discuss the solubility of gases such as carbon dioxide in water.

Ask for the reasons for the nonmiscibility of substances such as water and gasoline, water and cooking oil, and water and baby oil.

Ask whether substances such as salt and sugar will dissolve or not in alcohol or gasoline. Demonstrate this experiment and find out the answers.

The solubility of a solid or a liquid in water depends on several factors. It depends on the nature of water, the nature of chemical bonding in the solid or liquid, and their mutual compatibilities. Only similar substances dissolve in similar solvents. For example sugar and water are similar. A water molecule has partial electrical charges in its atoms. The sugar molecule also has partial electrical charges in its atoms. The partial positive charge on the hydrogen atom of the water molecule can attract the partial negative charge on the oxygen atom in the sugar molecule. The partial negative charge on the oxygen atom in the water molecule can attract the partial positive charge on the hydrogen atom in the sugar molecule. These attractive forces dominate and will result in the dissolution of sugar in water.

Similar reasoning explains the dissolution of salt and water. However, salt is an ionic solid, with a +1 electrical charge on the sodium atom and a −1 electrical charge on the chlorine atom instead of partial electrical charges.

A substance such as gasoline has practically no partial electrical charges and therefore, is not attracted to the partial electrical charges on the atoms of the water molecule. Therefore, gasoline and water do not mix. Substances such as ethanol or ethyl alcohol have one part of the molecule which is like gasoline and another part of the molecule which is like water. The attraction of the latter part with water dominates and therefore, alcohol is soluble in water. Substances such as baby oil and cooking oil also have two parts. However, the major portions of their molecules are similar to gasoline and therefore, these oils do not mix with water. The iodine used in this experiment is similar to gasoline in that it has no partial electrical charges in the molecule. Therefore, it does not dissolve in water. Instead it dissolves in a similar substance such as baby oil. The solid food colors have several atoms with partial electrical charges and their attractions with water dominate so that they are soluble in water.

The food colors are not true molecules. To be a true molecule, the substance must have covalent bonding only (sharing of electrons) between their atoms. The commonly used food colors are ionic substances because they have positive sodium ions and negative ions at some ends of the huge molecule. However, the food colors have been extensively used by many investigators as true molecules. This is partly justified because the charges are negligible compared to the huge size of the molecules.

TEACHER'S SECTION
ESCAPING OF MOLECULES OF A SOLID WITH THE HELP OF WATER

GOALS AND OBJECTIVES:

1. To determine whether molecules can escape from a paper towel with the help of water.

2. To learn that when a solid is dissolved in water, the size of the particles becomes very small.

DISCUSSION BEFORE THE LABORATORY ACTIVITY:

It is very important to stress the need for wearing an apron and safety goggles.

Discuss the relative motion of molecules in a solid, liquid, and gas.

Discuss the sizes of molecules.

Ask the students to suggest ways of observing molecular motion in a solid, in a liquid and in a gas.

Ask the students to predict the best way of storing a perfume and how to check out this prediction.

Ask the students to explain how we get the smell of a perfume when we open a bottle of perfume.

Ask the students to suggest a way of allowing sugar to pass through a coffee filter.

Ask the students to suggest an experiment using water to find out whether the tiny holes in a rubber balloon are bigger or smaller than the tiny holes in a paper towel. (The water should remain in the balloon and the water should drip through the paper towel.)

TEACHER'S SECTION
ESCAPING OF MOLECULES OF A SOLID FROM DIFFERENT CONTAINERS

GOALS AND OBJECTIVES:

1. To determine whether molecules of a solid can escape from various types of containers with the help of water.

2. To learn that when a solid is dissolved in water, the size of the particles becomes very small.

3. To learn that small particles, made by dissolving solids in water, can escape only from some containers.

4. To learn that different containers made from different materials have holes of different sizes.

DISCUSSION BEFORE THE LABORATORY ACTIVITY:

It is very important to stress the need for wearing an apron and safety goggles.

Discuss with students the relative motion of molecules in a solid, liquid, and gas.

Discuss the proximity between molecules in solids, liquids, and gases.

Discuss the sizes of molecules.

Challenge students to suggest ways of finding out the relative sizes of the holes in different containers made of different materials.

TEACHER'S SECTION
OXYGEN IN THE AIR

GOALS AND OBJECTIVES:

1. To test for oxygen in the air.//
2. To learn that oxygen supports combustion.
3. To learn that oxygen does not burn by itself.

DISCUSSION BEFORE THE LABORATORY ACTIVITY:

Stress the need to wear an apron and safety goggles.

Discuss the need to handle lighted candles carefully.

Discuss the reasons for not playing with matches.

Discuss the need for oxygen in the air.

Discuss some of the physical and chemical properties of oxygen.

Ask students to name substances that react with oxygen in the air. (A cut apple, for example, changes color due to its reaction with the oxygen in the air).

Quite often, this experiment has been described as a way of estimating the amount of oxygen in the air. This is wrong. This experiment can only show the presence of oxygen and the consumption of oxygen. It cannot measure the amount of oxygen present. An accidental combination of several compensating errors contribute to an observed reduction in volume of air of about 20%. For every molecule of oxygen consumed, a molecule of carbon dioxide may be formed. This cannot produce a reduction in pressure unless the carbon dioxide is removed by reacting with substances such as lime water or a solution of barium hydroxide. Carbon dioxide cannot be removed completely by dissolving it in water because it dissolves at a very low rate. The incomplete oxidation of the candle that produces carbon monoxide and carbon are not included. Oxygen also combines with the hydrogen from the hydrocarbons of the candle and produces water vapor. Water vapor can condense and cause a reduction in the pressure inside. If the bottle is not placed over the candle as soon as it is lit, the air surrounding the flame will get hot and expand. When it cools, this will cause a reduction in pressure inside the bottle. Also some gases from the system will be lost because of the expansion of the gases from heating by the candle. This happens if there is a considerable delay in placing the bottle over the lighted candle.

Gas molecules move very fast. For example, the speed of nitrogen molecules at 25°C is about 515 meters per second. The speed of a minimum hurricane wind is 75 miles per hour or about 34 meters per second. When molecules are moving, they collide with each other. A molecule collides with other molecules about 5 billion times a second. The average distance traveled by a molecule between collisions is extremely small, about 200 times the diameter of the molecule. Bigger molecules move slower than smaller ones.

TEACHER'S SECTION
CARBON DIOXIDE IN AIR

GOALS AND OBJECTIVES:

1. To test for the presence of carbon dioxide in the air.

2. To learn that carbon dioxide reacts with calcium hydroxide solution and produces an insoluble substance (a precipitate) called calcium carbonate.

3. To learn that calcium carbonate reacts with carbon dioxide in water and produces a soluble substance called calcium bicarbonate or calcium hydrogen carbonate.

DISCUSSION BEFORE THE LABORATORY ACTIVITY:

It is very important to stress the need for wearing an apron and safety goggles.

Discuss the need for carbon dioxide in the air.

Discuss photosynthesis briefly.

Discuss how carbon dioxide is produced from burning organic substances.

Discuss some of the physical and chemical properties of carbon dioxide.

Discuss some of the harmful effects (weather changes) from excess carbon dioxide in the atmosphere.

The chemical reactions are:

$$CaO + H_2O \longrightarrow Ca(OH)_2$$

Calcium oxide + water ⟶ Calcium hydroxide (lime water)

$$Ca(OH)_2 + CO_2 \longrightarrow CaCO_3 + H_2O$$

Calcium hydroxide + Carbon dioxide ⟶ Calcium carbonate + water

$$2\ CaCO_3 + H_2O + CO_2 \longrightarrow 2\ Ca(HCO_3)_2$$

Calcium hydroxide + Water + Carbon dioxide ⟶ Calcium bicarbonate or calcium hydrogen carbonate

A solution of barium hydroxide (prepared by dissolving a tablespoon of $Ba(OH)_2 \cdot 8H_2O$ in about 500 mL of water) is much more sensitive to carbon dioxide than lime water. It can detect much smaller quantities of carbon dioxide in the air. However barium salts are toxic and must be handled with extreme care. The barium carbonate formed is much less soluble in water than calcium carbonate.

Lime water is made by putting a tablespoon of calcium oxide into about 500 mL of water and shaking several times. Allow the excess calcium oxide to settle down. This may take about 24 hours. So prepare the lime water a day or two earlier. Use only the clear solution for your experiments.

Gas molecules move very fast. For example, the speed of nitrogen molecules at 25°C is about 515 meters per second. The speed of a minimum hurricane wind is 75 miles per hour or about 34 meters per second. When molecules are moving, they collide with each other. A molecule collides with other molecules about 5 billion times a second. The average distance traveled by a molecule between collisions is extremely small, about 200 times the diameter of the molecule. Bigger molecules move slower than smaller ones.

TEACHER'S SECTION
CHEMICAL CHANGES

GOALS AND OBJECTIVES:

1. To investigate some characteristics of chemical changes.
2. To learn that a chemical change changes the composition of the substance.
3. To learn that formation of gas bubbles is an indication of a chemical change.
4. To learn that a color change is an indication of a chemical change.
5. To learn that the disappearance of color is an indication of a chemical change.

DISCUSSION BEFORE THE LABORATORY ACTIVITY:

It is very important to stress the need for wearing an apron and safety goggles.

Caution students to handle hydrogen peroxide and clorox bleach with extreme care.

Discuss the properties of matter.

Discuss physical properties and chemical properties. Ask students to give some examples of both of these properties.

Remind students that a physical change does not change the composition of the substance. A chemical change changes the composition of the substance.

Ask students to suggest different ways of recognizing chemical changes.

Remind students of the need for extreme care while investigating chemical changes. Explosion of dynamite, fireworks, and fires are all due to chemical changes.

Formation of a gas on mixing two solids or two liquids is an indication of a chemical change. This should not be confused with the formation of bubbles on shaking a soda bottle.

Formation of a color on mixing two substances is an indication of a chemical change. However, this should not be confused with the formation of a new color by mixing two or more appropriate dyes.

The reaction between oxygen in the air and a slice of apple is an example of an oxidation–reduction or redox reaction. It is not a simple reaction, even though it is easy to do the experiment and observe the results.

Oxidation is the process of losing an electron. Reduction is the process of gaining an electron. Oxidizing agents help to oxidize and in that process they get reduced. Reducing agents help to reduce and in that process they get oxidized. Oxygen gains electrons and oxygen is an oxidizing agent in this reaction. An organic compound present in the slice of an apple loses electrons and is a reducing agent in this reaction.

The reaction between clorox bleach and green food color is also an example of an oxidation–reduction reaction. This is also not a simple one, even though it is easy to do the experiment and observe the results. The green food color is a mixture of yellow and blue food dyes. The yellow dye is bleached more easily than the blue dye. Therefore, it is possible to observe a change from green to blue to colorless on adding drops of clorox to the green food color. Thus both a color change and a color disappearance can be observed.

The clorox acts as an oxidizing agent and the dyes in the green food color act as a reducing agent in this reaction.

Baking soda is the same as sodium bicarbonate or sodium hydrogen carbonate, $NaHCO_3$. Vinegar is the same as a solution of 5% acetic acid, CH_3COOH. The reaction between them produces a salt, sodium acetate, CH_3COONa, water, and a gas, carbon dioxide, CO_2. The formation of the salt and water cannot be seen. However, the formation of carbon dioxide gas as bubbles can be observed.

$$NaHCO_3 + CH_3COOH \longrightarrow CH_3COONa + CO_2 + H_2O$$

The enzyme, catalase, present in the slice of an apple serves as a catalyst in the decomposition of hydrogen peroxide. Catalysts are substances that speed up chemical reactions. Catalysts remain the same before and after the reaction. Hydrogen peroxide decomposes to produce water and oxygen. The formation of water cannot be seen. However, the formation of oxygen bubbles can be observed.

$$2\ H_2O_2 \longrightarrow 2\ H_2O + O_2$$

This decomposition reaction is also an oxidation–reduction reaction. It is an example of an auto oxidation–reduction reaction. One molecule of hydrogen peroxide acts as an oxidizing agent and another molecule acts as a reducing agent.

TEACHER'S SECTION
WATER, SALT–WATER, AND SUGAR–WATER

GOALS AND OBJECTIVES:

1. To identify pure water, salt solution, and sugar solution without tasting them.
2. To learn that salt solution conducts electricity.
3. To learn that sugar solution does not conduct electricity.
4. To learn that sugar decomposes easily.
5. To learn that salt does not decompose easily.

DISCUSSION BEFORE THE LABORATORY ACTIVITY:

It is very important to stress the need to wear an apron and safety goggles.

Discuss the need for handling the conductivity apparatus with extreme care. Be sure your hands are dry when you connect the plug to the electrical outlet. Do not touch the electrodes with bare hands.

Discuss some of the characteristics of water. Ask the students to suggest methods of identifying water without tasting.

Discuss some of the characteristics of common salt. Discuss solubility of salt in water. Ask the students to suggest methods of identifying salt and salt solution without tasting. Discuss the commercial production of salt. Discuss some health hazards of eating too much salt. Ask the students to name some foods that taste very salty.

Discuss some of the characteristics of sugar. Discuss the solubility of sugar in water. Ask the students to predict the comparative solubility of salt and sugar in water. Check it out by doing an experiment. Discuss the commercial production of sugar. Discuss the problem of tooth decay from eating too many sweets. Ask them to name some foods that tasted too sweet. Discuss substitutes for sugar. Ask them to suggest methods of identifying sugar and sugar solution without tasting.

The solubility of common salt in 100 grams of water is 35.7 grams at $0°C$ and 39.12 grams at $100°C$. The solubility of sucrose in 100 grams of water is low in cold water and high in hot water.

The melting point of salt is 801°C and the boiling point of salt is 1413°C. Sucrose (cane sugar) melts at 185°C and decomposes before boiling.

The formula for salt is NaCl. This is called sodium chloride. The metal, sodium, and the nonmetal, chlorine, chemically combined to give an ionic compound, NaCl. During the chemical combination or bonding, the sodium loses an electron and chlorine gains an electron. When an ionic compound is formed, there is a transfer of an electron from one element to another element. In that process ions are formed. The ionic compound NaCl has sodium ions with a +1 charge, Na^+, and chloride ions with a -1 charge, Cl. The salt is really Na^+Cl^-. Since the opposite charges cancel each other, the salt is neutral and we write it as NaCl. However, the salt dissociates in water to give freely moving sodium ions and chloride ions. It is these freely moving ions that conduct the electricity.

The formula for sucrose is $C_{12}H_{22}O_{11}$. This molecule is formed by the sharing of electrons. Compounds in which the molecules are formed by the sharing of electrons are called covalent compounds. When the sugar dissolves in water, it does not dissociate into free ions. You need free ions to conduct electricity in solution.

Pure water does not conduct electricity. However, you can get electrical shocks from passage of electricity through household water, because it contains traces of many ionic compounds.

Prepare a solution in bottle A by dissolving 50 grams of salt in 1000 mL water. Prepare a solution in bottle B by dissolving 50 grams of sugar in 1000 mL water. Bottle C contains 1000 mL water. Sometimes, the solutions prepared from common salt and sugar may be a little cloudy. Prepare the solutions a day early and decant the clear solutions into similarly labeled bottles. Or filter the solutions. It is important that all three liquids in the three bottles look the same.

TEACHER'S SECTION
MAGIC OF CHANGING COLORS ON TALKING

GOALS AND OBJECTIVES:

1. To study the color changes induced by carbon dioxide in the exhaled air.
2. To learn that exhaled air contains more carbon dioxide than inhaled air.
3. To learn that carbon dioxide produces an acid when it is dissolved in water.
4. To learn that the colors of certain dyes are quite sensitive to the amount of acid in solution.
5. To learn that the presence of carbon dioxide can be tested using certain dyes instead of lime water.

DISCUSSION BEFORE THE LABORATORY ACTIVITY:

It is very important to stress the need to wear an apron and safety goggles.

Discuss the composition of air.

Discuss breathing. Discuss the differences between inhaled air and exhaled air. Ask the students to suggest ways of checking these differences.

Discuss carbonated beverages.

Discuss the need for carbon dioxide in the air.

Discuss briefly photosynthesis.

Discuss how carbon dioxide is produced from burning organic substances.

Discuss some of the harmful effects (weather changes) from excess carbon dioxide in the atmosphere.

Discuss some of the physical and chemical properties of carbon dioxide.

Discuss the formation of carbonic acid by the reaction of carbon dioxide and water.

The chemical reactions are:

$$CO_2 + H_2O \rightleftharpoons H_2CO_3$$

Carbon dioxide + water $\rightleftharpoons$ carbonic acid

$$H_2CO_3 \rightleftharpoons H^+ + HCO_3^-$$

Carbonic acid $\rightleftharpoons$ hydrogen ion + hydrogen carbonate ion or bicarbonate ion

$$HCO_3^- \rightleftharpoons H^+ + CO_3^{2-}$$

Hydrogen carbonate ion $\rightleftharpoons$ hydrogen ion + carbonate ion

The sign, $\rightleftharpoons$, indicates that the reaction is taking place in both the directions.

The hydrogen ion in water reacts with a water molecule to produce a hydronium ion. There are no free H^+ ions in water. They are always in the form of hydronium ions. Chemists often write hydronium ion as $H^+(aq)$ instead of H_3O^+. The short form for aqueous (means in water) is aq.

$$H^+ + H_2O \longrightarrow H_3O^+$$

Hydrogen ion + water $\longrightarrow$ hydronium ion

The amount of acid in water is determined by the amount of hydrogen ions, $H^+(aq)$ or hydronium ions, H_3O^+. The greater the number of hydronium ions, the greater the acidity.

A basic solution contains more hydroxide ions, OH^-, than hydronium ions.

To indicate the presence of very small amounts of acids and bases, chemists often use a pH scale. The pH scale is from 0 to 14 at 25°C. A pH of 0 indicates maximum acid concentration and a pH of 14 indicates minimum acid concentration. A pH of 7 is neutral. Solutions with pH less than 7 have more acid than base. Solutions with pH greater than 7 have more base than acid. When the pH value changes by one unit, the acid concentration changes by a factor of 10. For example, pH 4 is 10 times more acidic than pH 5. A pH of 3 is 100 times more acidic than pH 5. A pH of 13 is 10 times more basic than pH 12.

The extent of hydronium ions in water due to carbon dioxide is determined by the solubility of carbon dioxide and its reactions with water.

The chemical structures of phenol red and bromothymol blue in acid solutions and in base solutions are given at the beginning of the book, under the title, "Chemical Structures".

Phenol red changes from the red color in basic solutions to the yellow color in acidic solutions at a pH range of 8.0 – 6.6.

Bromothymol blue changes from the blue color in basic solutions to the yellow color in acidic solutions at a pH range of 7.6 – 6.0

TEACHER'S SECTION
ENDOTHERMIC PROCESSES

GOALS AND OBJECTIVES:

1. To determine whether heat energy is absorbed in some processes.

2. To learn that energy absorbed by chemicals is stored as potential energy.

3. To learn that energy absorption results in a lowering of temperature.

DISCUSSION BEFORE THE LABORATORY ACTIVITY:

It is very important to stress the need for wearing an apron and safety goggles.

Discuss heat energy and the flow of heat energy.

Discuss the meaning of temperature.

Discuss exothermic and endothermic processes.

One calorie is the amount of heat needed to raise the temperature of 1 gram of water by 1° (Celsius). The exact definition of calorie specifies the temperature change of the water from 14.5 to 15.5 °C.

This calorie is different from the food calorie. One food calorie = 1000 calories = 1 kilocalorie or 1 kcal.

Nowadays scientists use another unit, joule, for energy. One joule = 4.184 calories. One joule is the energy which is equal to the work done by a force of one Newton acting over a distance of one meter.

Specific heat is the heat required to raise the temperature of 1 gram of any substance by 1°C.

The heat of fusion of ice is 80 calories per gram. This means that when one gram of water at 0°C freezes into ice, it releases 80 calories of heat. This is not a small amount. Most substances have much less heat of fusion. Even though ice seems to melt easily, it is one of the most difficult substances to melt. One gram of iron can be melted more easily at its melting point with less heat.

Temperature is a measure of the average kinetic energy of the molecules. Kinetic energy is the energy from the motion of molecules. There are three common temperature scales, Fahrenheit (F), Celsius (C), and Kelvin (K).

The fixed points on the Fahrenheit and Celsius thermometers are based on the melting point of ice and the boiling point of water. It is assumed that ice melts at 32°F or 0°C and water boils at 212°F or 100°C at one atmospheric pressure. The more meaningful Kelvin scale is based on the lowest temperature possible to attain. The lowest possible temperature is –273°C. This temperature is called absolute zero or zero Kelvin. Scientists have succeeded in reaching very close to absolute zero but not absolute zero itself. According to this Kelvin scale (also called absolute scale), ice melts at 273 K and water boils at 373 K.

Energy changes in the form of heat, light, and electricity can take place during a chemical reaction. Electrical energy is produced in batteries from chemical reactions. In lightsticks, light energy is produced from chemical reactions. Absorption or release of heat energy is the most common process in most chemical reactions.

In an exothermic reaction or process, energy is released as heat. This energy comes from the stored energy in the substance. The energy released in an exothermic reaction is usually expressed as a negative number. This does not mean negative heat. It only means that much energy is released from the system into the surroundings.

In an endothermic reaction or process, energy is absorbed and stored as potential energy. The energy absorbed in an endothermic reaction is usually expressed as a positive number.

Common sense suggests that the temperature should increase on absorption of heat energy. However, in an endothermic reaction, the energy is absorbed from the system and is stored as potential energy. A thermometer cannot measure potential energy. Temperature, measured by a thermometer, is a measure of the average kinetic energy of the molecules.

The formulas for the different substances are: sodium chloride, $NaCl$; ammonium chloride, NH_4Cl; ammonium nitrate, NH_4NO_3; and potassium iodide, KI. The heat of solution of sodium chloride is 16 calories/gram. The heat of solution of ammonium chloride is 66 calories/gram. The heat of solution of ammonium nitrate is 78 calories/gram. The heat of solution of potassium iodide is 29 calories/gram. Heat of solution is the heat absorbed or released when a known amount of the solute is dissolved in a large amount of water so that the solution is very dilute. Judged by the above data, the dissolution of ammonium nitrate in water must be the most endothermic and that of sodium chloride must be the least endothermic.

TEACHER'S SECTION
EXOTHERMIC PROCESSES

GOALS AND OBJECTIVES:

1. To determine whether energy is released as heat in some processes.

2. To learn that energy released by chemicals comes from their stored potential energy.

3. To learn that energy release results in an increase in temperature.

DISCUSSION BEFORE THE LABORATORY ACTIVITY:

It is very important to stress the need to wear an apron and safety goggles.

Discuss heat energy and the flow of heat energy.

Discuss the meaning of temperature. Temperature is a measure of the average kinetic energy of the molecules. Kinetic energy is the energy from the motion of molecules. There are three common temperature scales, Fahrenheit (F), Celsius (C), and Kelvin (K). The fixed points on the Fahrenheit and Celsius thermometers are based on the melting point of ice and the boiling point of water. It is assumed that ice melts at 32°F or 0°C and water boils at 212°F or 100°C at one atmospheric pressure. The more meaningful Kelvin scale is based on the lowest temperature possible to attain. The lowest possible temperature is –273°C. This temperature is called absolute zero or zero Kelvin. Scientists have succeeded in reaching very close to absolute zero but not absolute zero itself. According to this Kelvin scale (also called absolute scale), ice melts at 273 K and water boils at 373 K.

One calorie is the amount of heat needed to raise the temperature of 1 gram of water by 1° (Celsius). The exact definition of calorie specifies the temperature change of the water from 14.5 to 15.5 °C. This calorie is different from the food calorie. One food calorie = 1000 calories = 1 kilocalorie or 1 kcal.

Nowadays scientists use another unit, joule, for energy. One joule = 4.184 calories. One joule is the energy which is equal to the work done by a force of one Newton acting over a distance of one meter.

Specific heat is the heat required to raise the temperature of 1 gram of any substance by 1°C.

The heat of fusion of ice is 80 calories per gram. This means that when one gram of water at 0°C freezes into ice, it releases 80 calories of heat. This is not a small amount. Most substances have much less heat of fusion. Even though ice seems to melt easily, it is one of the most difficult substances to melt. One gram of iron can be melted more easily at its melting point with less heat.

Discuss exothermic and endothermic processes.

Energy changes in the form of heat, light, and electricity can take place during a chemical reaction. Electrical energy is produced in batteries through chemical reactions. In lightsticks, light energy is produced by chemical reactions. Absorption or release of heat energy is the most common process in most chemical reactions.

In an exothermic reaction or process, energy is released as heat. This energy comes from the stored energy in the substances. The energy released in an exothermic reaction is usually expressed as a negative number. This does not mean negative heat. It only means that much energy is released from the system into the surroundings.

In an endothermic reaction or process, energy is absorbed and stored as potential energy. The energy absorbed in an endothermic reaction is usually expressed as a positive number.

The formulas for the substances used here are: anhydrous sodium carbonate, Na_2CO_3; anhydrous sodium sulfate, Na_2SO_4; anhydrous calcium chloride, $CaCl_2$; and ethyl alcohol (ethanol), C_2H_5OH.

The heat of solution values for the different substances are: anhydrous sodium carbonate, −53 calories/gram; anhydrous sodium sulfate, −3.2 calories/gram; anhydrous calcium chloride, −157 calories/gram; and ethyl alcohol, −53 calories/gram. Heat of solution is the heat absorbed or released when a known amount of the solute is dissolved in a large amount of water so that the solution is very dilute. Judged by the above data, the dissolution of anhydrous calcium chloride in water must be the most exothermic and that of anhydrous sodium sulfate must be the least exothermic.

It is important that the substances at the start are dry. Otherwise, there will be substantial changes in the amounts of heat released. For example the heat of solution of sodium sulfate monohydrate, $Na_2SO_4 \cdot H_2O$, is +12 calories/gram and that of sodium sulfate decahydrate, $Na_2SO_4 \cdot 10H_2O$, is +58 calories/gram.

There may be some difficulty dissolving the anhydrous sodium carbonate and anhydrous sodium sulfate. If this happens, it is better to have an initial temperature for the water slightly higher than room temperature.

TEACHER'S SECTION
MAXIMUM AMOUNT OF GAS FROM THE SHELL OF AN EGG

GOALS AND OBJECTIVES:

1. To determine the maximum amount of gas that can be collected by decomposing an egg shell.
2. To learn that the shell of an egg reacts with vinegar.
3. To learn that a gas is produced when an egg shell is decomposed using vinegar.

DISCUSSION BEFORE THE LABORATORY ACTIVITY:

It is very important to stress the need for wearing an apron and safety goggles.

Caution students to be careful in handling raw eggs.

Discuss chemical reactions.

Discuss the composition of the shell of an egg.

Discuss ways of removing the shell from a raw egg.

The shell of an egg is mostly calcium carbonate. All carbonates decompose by reacting with acids. Vinegar is a solution of 5% acetic acid, CH_3COOH. All acids produce hydrogen ions, $H^+(aq)$, or hydronium ions, H_3O^+, in water. Vinegar reacts with the shell and produces soluble calcium acetate, water and carbon dioxide. The reaction is:

$$CaCO_3 + 2\ CH_3COOH \longrightarrow Ca(OOCCH_3)_2 + H_2O + CO_2$$

Calcium Carbonate + Acetic Acid ⟶ Calcium Acetate + Water + Carbon Dioxode

The reaction is rather slow. Also the formation of bubbles on the egg shell slows down the reaction. Occasional swirling of the flask to release the bubbles sticking to the egg shell will speed the decomposition process. The amount of gas collected depends on the mass of the egg shell. To make it more quantitative, one should weigh the shell before allowing it to react with the vinegar. One gram of egg shell should produce about 240 mL of carbon dioxide gas at a temperature of 25°C and pressure of one atmosphere.

The results of the experiment can be used to find the egg shell with the least mass without actually weighing them.

TEACHER'S SECTION
CARBON DIOXIDE GAS FROM THE SHELL OF AN EGG

GOALS AND OBJECTIVES:

1. To identify the gas that can be collected by decomposing an egg shell and other carbonates.
2. To learn that an egg shell reacts with vinegar and produces a gas.
3. To learn that the gas produced by decomposing an egg shell is carbon dioxide.
4. To learn two or three tests for carbon dioxide.

DISCUSSION BEFORE THE LABORATORY ACTIVITY:

It is very important to stress the need for wearing an apron and safety goggles.

Caution students to be careful when they handle raw eggs.

Discuss chemical reactions.

Discuss the composition of the shell of an egg.

Discuss ways of removing the shell from a raw egg.

The shell of an egg is mostly calcium carbonate. All carbonates decompose by reacting with acids. Vinegar is a solution of 5% acetic acid, CH_3COOH. All acids produce hydrogen ions, $H^+(aq)$, or hydronium ions, H_3O^+, in water. Vinegar reacts with the shell and produces soluble calcium acetate, water, and carbon dioxide. The reaction is:

$$CaCO_3 + 2\ CH_3COOH \longrightarrow Ca(OOCCH_3)_2 + H_2O + CO_2$$

The reaction is rather slow. Also the formation of bubbles on the egg shell slows down the reaction. Occasional swirl the flask to release the bubbles sticking to the egg shell to speed the decomposition process.

Carbon dioxide is not a supporter of combustion. Therefore it will extinguish the flame from a candle.

Carbon dioxide also reacts with lime water and produces an insoluble white precipitate of calcium carbonate. This precipitate dissolves on continued reaction with excess carbon dioxide and water by producing soluble calcium hydrogen carbonate.

The chemical reactions are:

$$CaO + H_2O \longrightarrow Ca(OH)_2$$

Calcium oxide + Water ⟶ Calcium hydroxide (lime water)

$$Ca(OH)_2 + CO_2 \longrightarrow CaCO_3 + H_2O$$

Calcium hydroxide + Carbon dioxide ⟶ Calcium carbonate + water

$$2\,CaCO_3 + H_2O + CO_2 \longrightarrow 2\,Ca(HCO_3)_2$$

Calcium hydroxide + Water + Carbon dioxide ⟶ Calcium bicarbonate or calcium hydrogen carbonate

Carbon dioxide reacts with water and produces carbonic acid. This carbonic acid ionizes in water to produce hydrogen ions, $H^+(aq)$, or hydronium ions, H_3O^+.

The chemical reactions are:

$$CO_2 + H_2O \rightleftharpoons H_2CO_3$$

Carbon dioxide + water ⇌ carbonic acid

$$H_2CO_3 \rightleftharpoons H^+ + HCO_3^-$$

Carbonic acid ⇌ hydrogen ion + hydrogen carbonate ion or bicarbonate ion

$$HCO_3^- \rightleftharpoons H^+ + CO_3^{2-}$$

Hydrogen carbonate ion ⇌ hydrogen ion + carbonate ion

The sign, ⇌, indicates that the reaction is taking place in both the directions.

The hydrogen ion in water reacts with a water molecule to produce a hydronium ion. There are no free H^+ ions in water. They are always in the form of hydronium ions. Chemists often write hydronium ion as $H^+(aq)$ instead of H_3O^+. The short form for aqueous (means in water) is aq.

$$H^+ + H_2O \longrightarrow H_3O^+$$

Hydrogen ion + water ⟶ hydronium ion

The amount of acid in water is decided by the amount of hydrogen ions, $H^+(aq)$ or hydronium ions, H_3O^+. The greater the number of hydronium ions, the greater the acidity.

Blue litmus paper turns red in an acid solution.

It is also possible to test this acid produced using other indicators such as phenol red and bromothymol blue.

TEACHER'S SECTION
BLEACHING OF FOOD COLORS

GOALS AND OBJECTIVES:

1. To determine which food color is most easily bleached by clorox.
2. To learn that bleaching of color is a chemical reaction.
3. To learn that a chemical change changes the composition of the substance.
4. To learn that a color change is an indication of a chemical change.
5. To learn that the disappearance of color is an indication of a chemical change.

DISCUSSION BEFORE THE LABORATORY ACTIVITY:

It is very important to stress the need to wear an apron and safety goggles.

Caution students to handle clorox bleach with extreme care.

Discuss color and the color of the food dyes.

Find the F D & C No. of the food colors from the packages.

Ask students to produce other colors using red, blue and yellow dyes. Green is a mixture of yellow and blue dyes. Demonstrate this by mixing one drop of blue and one drop of yellow in a clear plastic cup. Add about 50 mL water to this and observe. Show the purple color by mixing one drop of blue and one drop of red in a clear plastic cup. Add about 50 mL water to this and observe.

Show the complicated structures of the dyes to the students. These structures are not easily available from text books. The chemical structures of the most commonly used food dyes are given at the beginning of the book, under the title, "Chemical Structures".

Discuss physical properties and chemical properties. Ask the students to give some examples of these properties.

Remind students that a physical change does not change the composition of the substance. A chemical change changes the composition of the substance.

Ask students to suggest different ways of recognizing chemical changes.

Remind students the need for extreme care while investigating chemical changes. Explosion of dynamite, fireworks, and fires are all due to chemical changes.

Formation of a color when two substances are mix is an indication of a chemical change. However, this should not be confused with the formation of a new color by mixing two or more appropriate dyes.

Oxidation is the process of losing an electron. Reduction is the process of gaining an electron. Oxidizing agents help to oxidize and in that process they get reduced. Reducing agents help to reduce and in that process they get oxidized.

The reaction between clorox bleach and food color is an example of an oxidation–reduction reaction. This is not a simple one, even though it is easy to do the experiment and observe the results.

The green food color is a mixture of yellow and blue food dyes. The yellow dye is bleached more easily than the blue dye. Therefore, it is possible to observe a change from green to blue to colorless when you add drops of clorox to the green food color. Thus both a color change and a color disappearance can be observed.

The clorox acts as an oxidizing agent and the dye in the food color acts as a reducing agent in this reaction.

General Science Basic Activities

by

Eugene Kutscher

TEACHER'S GUIDE
A MILLION STARS

GOAL: To develop student understanding of large numbers and the relationships between them.

STUDENT OBJECTIVES:
- To develop a concrete appreciation of the concept of one million objects.
- To relate large numbers to everyday activities such as the passage of time or the counting of money.

PRELAB DISCUSSION: Students can begin this experiment directly, or they may be encouraged to discuss areas in which large numbers commonly occur, such as in astronomy and the federal budget. The idea is to have fun while developing the desired concepts. Leave the stars up. If possible, do the activity in the hallways. The entire school will be talking about it! If the students have sufficient math background, this activity is a fine starting point for an introduction to scientific notation.

GUIDE TO THE INVESTIGATION:

Students will need masking tape, a clock or timer and a 200 page supply of the enclosed page which is completely covered with stars. The page may be reproduced directly or used as a model for reproduction. The following simple computer program, when directed to a printer, will generate a similar page:

```
10 FOR I = 1 TO 5000
20 PRINT "*";
30 NEXT I
```

This activity can be conducted by groups working individually or collectively. It may be extended, with posters and graphics, to a more formal school corridor display.

Students should complete the questions in the procedure section of their activity.

Students should be given an area in which to mount their stars. The walls of a classroom are an excellent choice. The ceiling may be used, but requires caution in the proper use of a ladder.

Discuss the results. Ask the students how many groups doing the same activity would be needed to display a billion stars. It is stimulating to the students to learn that astronomers have evidence that galaxies contain an average of 200 billion stars and that they know of ten billion galaxies!

VOCABULARY: galaxy, million, billion

ADDITIONAL RESOURCES: The following children's picture book is interesting to all age groups: *How Much is a Million?* (David M. Schwartz, Lothrop, Lee & Shepard, 1985)

TEACHER'S GUIDE
CAN YOU DESCRIBE THIS?

GOAL: To have students recognize the need for accuracy in description.

STUDENT OBJECTIVES:
- To distinguish between scientific and non-scientific description.
- To describe various objects as accurately as possible without the use of instruments.
- To learn which descriptions are quantitative and which are qualitative.

PRELAB DISCUSSION: In this experiment, a brief reading will be used as an introduction to an exciting and interesting activity.

In general, the trickier you are, the more fun you and your students will have. So be creative and have fun!

GUIDE TO THE INVESTIGATION:

The teacher can use these suggested objects (anything will work):

sneaker
can of corn or peas
statue or knickknack
fake pencil (see illustration)

Set of at least four, different size, but identical shape and design, boxes or canisters which nest into each other.

Take a piece of dowel as thick as an average pencil (red is ideal), cut it to a length of 12 cm, wrap a small piece of half-inch masking tape around the dowel about 1 cm from one end, color the same end so it looks like an eraser (leave it alone if it is red), sharpen the tip of the other end and color the point black using a pencil.

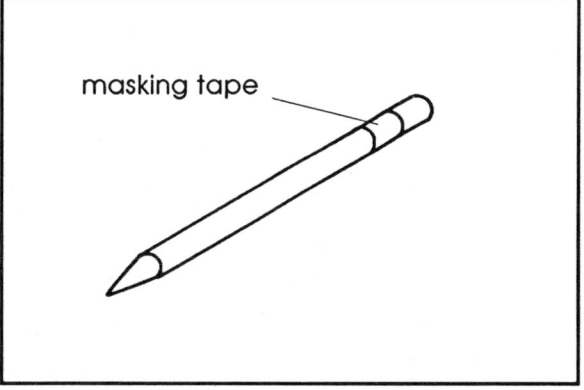

Fake pencil

Students will need a pencil and paper.

Have the students complete the reading activity. Discuss the results.

Tell the students that they are going to play a game. Discuss the procedures as listed. For each object, have a student go outside the door and wait down the hallway until you call for them. Keep all of the objects hidden. Bring out the first object. It is good to use an everyday, but not easily described object such as a sneaker.

The students must not use words like sneaker or descriptions of what you can do with the object. Only descriptive words of the object are allowed. For example, "two, 15 cm long, 1 cm thick, round, cloth, white objects" is acceptable, "laces" is not. Call on several volunteers to see if they are doing this properly. Hide the object, then allow them to finish.

Bring the student back into the room and have the student stand in front with you. Write the word "descriptions" on the chalkboard. Call on students, one at a time, to give a descriptive word or phrase of the object. Write it down. After each one, ask the student who is with you to guess what the object is. After about 10 clues, have the student sit. Bring out the object and comment on the accuracy of the clues.

Repeat this procedure for each object. However, when doing the fake pencil, do not show the object for very long. Students will think they know this one. When the student has guessed "pencil," (and they always do) show the object, put it away, and then say: "You know, I do not think that you guessed that one correctly." Bring out the object, explain what you have done, and pass it around. Be prepared for boos and hisses! Use this to explain the need for direct examination whenever possible.

Save the nesting boxes for last. Show the class a middle box. After the guessing, do not show the object; show the next one in line instead. Do not say anything. If someone says, "that is not it," remove the object and bring up the largest or smallest. Now everyone will say that is not it. Put the object back and bring out all of the boxes. Ask them to pick the right one. Use this to show the need for quantitative, as well as qualitative measurements.

VOCABULARY: density, qualitative, quantitative, state of matter

ADDITIONAL RESOURCES: Fake pencil aside, the importance of indirect evidence is great. For example, much of our knowledge of astronomy is based on indirect evidence. Use a book such as *Astronomy, Realm of the Universe*, Abell, Saunders, New York, 1989, to show how we find out how far away stars are and how hot they are. You may also wish to construct a black box experiment, using taped boxes with marbles, wood or other objects inside. Students can shake, turn, wave, tilt or weigh their boxes, but cannot look inside. They must write down their evidence and deduce the contents.

TEACHER'S GUIDE
THE BANANA RULER

GOAL: To develop student comprehension of the need for clearly understood worldwide standards of measurement.

STUDENT OBJECTIVES:
- To develop a personal standard of measurement and to attempt to communicate its meaning to others in the class.
- To use a standard measure of distance and to compare its ease of communication with that of the personal standard.

PRELAB DISCUSSION: The ability of scientists to communicate with each other rapidly and clearly is an important factor in the speed of advance of new discoveries and technologies. In this activity, the students will attempt to measure their desks using bananas and then other objects. Of course, any fruit or vegetable may be used. Bananas have the advantage of being immediately edible after the students are done with them. When asked to choose their own standard, many try silly things like chairs. As long as safety is kept in mind, this works quite well. Next, the students will attempt to communicate the measurements of their desks to each other. Finally, they will measure their desks with standard meter sticks or metric rulers and again try to communicate their results. The need for universally comprehended standards of measurement, such as those of the metric system, should be obvious. Students may work alone or in groups for this activity.

GUIDE TO THE INVESTIGATION:

Students will need:

> banana   metric ruler or meter stick
> other object of their choosing

Have the students conduct the activity. Review the results.

VOCABULARY: metric system, standard

ADDITIONAL RESOURCES: The history of measurement makes fascinating reading. The evolution of terms such as the English inch, foot, yard, mile and stone, the French metric system and the temperature scale of Fahrenheit may be assigned as extra credit reports. Ask the students: Why did these terms survive while those of the Egyptians and Greeks did not? A relatively easy reference for students is: Bendick, Jeanne, *How Much and How Many*, New York, McGraw-Hill,, 1960. Another is: Pine, Tillie S. and Levine, Joseph, *Measurements and How We Use Them*, New York, McGraw-Hill, 1977.

204 *Fundamentals of Science Activity Series*

TEACHER'S GUIDE
WHY DO WE NEED THERMOMETERS?

GOAL: To understand the need for a device to measure temperature.

STUDENT OBJECTIVES:
- To describe the limitations of human senses as temperature detectors.
- To learn the need for a temperature standard.
- To determine the freezing and boiling points of water in degrees Celsius.

PRELAB DISCUSSION: In this experiment, students will develop their own temperature scale and see the need for a standard scale. This activity is enhanced if students have done *The Banana Ruler* activity first. In the beginning of this activity, students will see that their skin detects relative, rather than absolute temperatures. Then they will determine the freezing and boiling points of water in their own units, as well as in Celsius units. Students should work in groups of two or more. Thermometers should be kept from the bottoms or sides of beakers. They may be suspended as indicated, or by using thermometer clamps or other reasonable methods.

BE SAFE! Goggles should be worn.

GUIDE TO THE INVESTIGATION:

Begin with a demonstration. You will need:

    3 Pyrex beakers, 500 ml   water
       or larger

BE SAFE! Before students come in, obtain some water at about 4 degrees Celsius, some at 20 degrees and heat some to 40 degrees. Put each in a beaker and fill it to the two-thirds level, as shown in student illustration 1. Ask the students if the hand is a good judge of temperature. Ask for two volunteers. **Warn them to be careful, as the liquid might be hot!** Ask one to put a hand in beaker A and the other hand in B. Which is warmer? Ask the second student to put hands in B and C. Ask the same question. Have the students reverse their assignments. Each will claim that B is both warmer **and** colder. Point out to the class that this does not permit easy communication among scientists or people in general. Therefore, numbers are needed.

Students will need:

Pyrex beaker, 250 ml or larger	water
hot plate	ice
unmarked thermometer	marking pen
Celsius thermometer	slotted cork
ring stand and clamp	goggles

BE SAFE! Remind students **to be careful with hot plates, and to turn them off as soon as they are done.**

Have the students carry out their portion of the activity.

VOCABULARY: standard, temperature

ADDITIONAL RESOURCES: The work of Fahrenheit, as well as the development of the metric system in France, are traced nicely in all major encyclopedias. Students also might be asked to look up the less known work of Galileo with thermometers.

TEACHER'S GUIDE
HOW MUCH SPACE DO YOU TAKE UP?

GOAL: To understand the concept of volume.

STUDENT OBJECTIVES:
- To define volume as the space occupied by something.
- To express volume in metric units.
- To determine the equivalence of the volume of a rectangular solid determined by formula or by counting cubes.
- To determine the volume of a small stone.
- To determine their own volume.

PRELAB DISCUSSION: This experiment is a great deal of fun, yet is successful at developing a thorough understanding of the concept of volume. Take pictures!! Students should work in groups of four. **Some students may not want to find their own volume in public. Do not make them! Be certain that all those who do are thoroughly dried after the activity.**

A good beginning is to place on a desk a liter bottle, a liter cube, a liter graduated cylinder and a liter Florence flask, if available. Ask which is bigger in volume or space, but do not say that they are liter containers. Let the class debate. Pour water from one to the other to demonstrate that they are equal. Again, you are presenting a case for the need for instruments to measure something. Reinforce this by placing a 10 ml cylinder on the desk. Ask how many are needed to fill the liter bottles. Few will guess more than 25.

When determining student volumes, ANY method of displacement is possible. If you do not wish to use water, use a substance that is denser than water as well as insoluble in it. Marbles work, as do baseballs. However, be certain to determine the air space between the marbles or baseballs:

personal volume = total volume - marble volume - air volume

The air volume is found by removing the person and adding a known amount of water until the marbles are just covered. If 4.22 l of water are added, and the total volume is 9.22 l, then the air volume is 4.22 l and the marble volume is 5.00 l.

GUIDE TO THE INVESTIGATION:

Begin with the above demonstration. You will need:

liter bottle	liter cylinder
liter cube	liter Florence flask
water	(optional)
	10 ml cylinder

Review the formula for the volume of a rectangular solid.

Students will need:

64 ml cubes (or sugar cubes)	small rock
graduated cylinder	water

The class will need:

barrel big enough for the students
water
bathing suits
towels
liter bottles
marking pen
step stool
chair

Have the students carry out their portion of the activity. Supervise the barrel activity closely.

VOCABULARY: volume

ADDITIONAL RESOURCES: The exploration of **volume** is developed in the text: Haber-Schaim, Uri, et. al., *Introductory Physical Science*, Englewood Cliffs, New Jersey, Prentice-Hall, Inc. (any edition).

TEACHER'S GUIDE
SINKING THE SODA

GOAL: To increase student understanding of the concept of density.

STUDENT OBJECTIVES:
- To determine the density of several objects.
- To visually determine the density of other objects.
- To explain sinking and floating cans of soda.

PRELAB DISCUSSION: This activity develops intense student understanding of the generally abstract concept of density. Each part may be done separately. In addition, the visual determination part may be done, alternatively, as a wonderful demonstration (use a liter cylinder if this is done). Students should know how to use a balance and should understand the terms mass, weight, and volume before they conduct this activity. A discussion of what density means also should precede this laboratory experience. For **Part 1**, the blocks should fit easily into the beaker, the rock should fit easily into the cylinder and blocks and rock should be within the capacity of the balance. Students should work in groups of two or more.

BE SAFE! Students should wear goggles.

GUIDE TO THE INVESTIGATION:

Students will need:

Part 1:

- balance
- metric ruler
- block of wood
  - (rectangular or cubic)
- block of metal
  - (rectangular or cubic)
- 600 ml or larger beaker
- water
- 100 ml graduated cylinder
- small rock

Part 2:

- materials from **Part 1**
- rubber stopper
- plastic checker or toy

cooking oil (density is about .90g / cc)
glycerine (glycerol) (density is about 1.25 g / cc)
corn syrup (density is about 1.40 g / cc)
goggles

Part 3:

can of Diet Coke or Diet Pepsi   fish tank or pail
can of same brand of regular cola   water
balance

Have the students conduct the investigation. Review the results.

VOCABULARY: density, mass, volume, weight

ADDITIONAL RESOURCES: The densities of many materials are listed in *The Handbook of Chemistry and Physics*.

TEACHER'S GUIDE
THE JELLO CELL

GOAL: To generate student comprehension of the structure and functions of the cell.

STUDENT OBJECTIVES:
- To construct a model of the cell, using Jello, to show the locations and relative sizes of the parts of the cell.
- To describe the functions of the major parts of cells.

PRELAB DISCUSSION: The study of the cell is central to the study of life and life functions. This exercise gives the student a joyous method of discovery while allowing great flexibility in terms of materials and concept levels which are used. Students should be guided with regard to the locations and relative sizes of the cell parts. Pictures or models are very helpful. The students may construct a generalized cell, a plant cell an animal cell or any other specific cell. Any number of organelles may be investigated, depending on the level of the group. It is suggested that students work in teams of two or more for this investigation. Different groups may do different cells for comparison or for extra credit. Each group may need several packages of Jello. Take pictures! When they are done they may eat the experiment!!

GUIDE TO THE INVESTIGATION:

Students will need:

> Jello mix and water
> bowl, or large waterproof plastic bag* or other container
> fruit of various sizes and shapes (other food may be used)
> glass or measuring cup
> stirring spoon
> aprons

The materials will have to be chilled in order to set.

* A large, clear, turkey microwaving bag is an example.

VOCABULARY: cell, cell membrane, cell wall, cytoplasm, nucleus, organelle, vacuole (and other organelles, as desired)

ADDITIONAL RESOURCES: Carolina Biological Supply sells models and diagrams of the cell. Pictures of cells and descriptions of the functions of their parts are in virtually all life science and biology books.

TEACHER'S GUIDE
ENZYMES AND YOU

GOAL: To develop student understanding of the role of enzymes in converting large and complex molecules into smaller, simpler ones.

STUDENT OBJECTIVES:
- To describe the properties of enzymes.
- To determine the effect of amylase on starch.
- To determine the action of pepsin on protein.

PRELAB DISCUSSION: This multiple activity includes testing for nutrients. Begin by describing a catalyst. Describe enzymes as catalysts in your body. Tell the students that, in fact, enzymes are required for virtually all chemical reactions in all living things!

BE SAFE! **Review procedures for heating, and for testing for starch and sugar. In addition, students should be reminded how to work with small amounts of strong acids and bases. If the students are not at a level sufficient for working with these materials, add them yourself or omit Part 2!**

BE SAFE! In the past, saliva was used in this experiment. **Given the nature of modern concerns over diseases, it is best to obtain amylase, the enzyme in saliva, from a biological supply house.** Plant amylase, called diastase, can be used as well.

BE SAFE! **Students should wear goggles and a laboratory apron.** They should work in groups of two or more.

GUIDE TO THE INVESTIGATION:

For **Part 1**, students will need:

   drying oven, or incubator, or anything that can be held at about body temperature (37 degrees Celsius) for 48 hours.

   hot plate
   balance
   3 Pyrex test tubes
   400 ml or larger pyrex beaker
   test tube rack or basket

   5 ml amylase (.5 %)
   Lugol's iodine solution
   Benedict's solution
   1 saltine cracker
   marking pencil

test tube holder or pot holder　　goggles
water　　laboratory apron
graduated cylinder

For **Part 2**, students will need:

drying oven　　1 hard boiled egg
balance　　water
4 test tubes　　paper towels
test tube rack or basket　　marking pencil
knife or scalpel　　goggles
10 ml pepsin solution　　laboratory apron
graduated cylinder
dropper bottle of hydrochloric acid (1 per class can be used)
dropper bottle of sodium hydroxide (1 per class can be used)

Have the students conduct the investigation. Review the results.

VOCABULARY: amylase, catalyst, enzyme, glucose, pepsin, protein, starch

ADDITIONAL RESOURCES: The preparation of a starch digesting enzyme may be found in simple form in *Principles of Science, Book Two, Activity-Centered Program Teacher's Guide*, Columbus, Ohio, Charles E. Merrill Publishing Co., 1979. A good reference for organic enzymes and their actions is: Stryer, L., *Biochemistry*, San Francisco, Freeman & Co., 1981.

TEACHER'S GUIDE
YOUR ROOTS: WHY YOU ARE YOU

GOAL: To have students understand the role of genetics in giving them the inheritable traits which they have.

STUDENT OBJECTIVES:
- To determine their own phenotypes and possible genotypes for two inherited traits.
- To prepare a family pedigree chart for those two traits.

PRELAB DISCUSSION: In this activity, two easily observable traits, inherited from single gene pairs, will be used to develop a family pedigree chart for the two characteristics. Students should be familiar with the concept of a gene and the idea that genes usually come in pairs; one from each parent. Concepts such as dominant and recessive, pure and hybrid should be discussed. It may be desirable, but is not necessary, to use the words genotype and phenotype. If time is a problem, either trait may be investigated on its own. Review the construction of a pedigree chart. A simple version is included in the student section of this activity. Remind students that the more relatives they can test, the more accurate their pedigree will be. Be sensitive to the needs and feelings of students who may not live with natural parents. Have several extra family descriptions for these traits prepared in advance for their use.

GUIDE TO THE INVESTIGATION:

Students will need:

    mirror (or a partner)
    PTC paper (from any scientific supply company)

Have the students conduct their investigation in the classroom and at home. Review the results. There is virtually no limit to the number of traits and concepts (blending, multiple gene inheritance, probability, sex-linked characteristics) which may be studied for supplemental work.

VOCABULARY: carrier, chromosome, dominant, gene, genotype, hybrid, pedigree, phenotype, pure, recessive, trait

ADDITIONAL RESOURCES: Local breeders usually will make available pedigree charts of dogs, cats, horses and prize farm animals.

TEACHER'S GUIDE
FOOD CHAINS AND FOOD WEBS

GOAL: To understand the balance of nature in food chains and webs.

STUDENT OBJECTIVES:
- To become the creatures in a food web.
- To discover the interrelationships in a typical food web.
- To discover the impact of a change in one organism in the food web on the entire web.
- To realize that pollution and habitat destruction cause such changes.

PRELAB DISCUSSION: This investigation develops the concepts of ecology in a simple and tangible way. It can be enhanced, effectively, by a study of a local ecosystem. It presents an excellent opportunity to introduce or reinforce the concept of pollution and its effects on local, national and global ecosystems. A similar opportunity is present for investigating the effects of habitat destruction, such as the worldwide reduction of rain forests. Take a picture of the food web!

GUIDE TO THE INVESTIGATION:

The class will need:

>   20 white cardboard cards, about 10 cm x 40 cm
>   dark marking pencil or crayon
>   colored yarn (one bright color will do, but many colors enhance the visual effect)
>   tape or string

Carry out the procedures with the students. Review the results.

VOCABULARY: balance of nature, community, consumer, ecology, food chain, food web, habitat, pollution, predator, prey, producer

ADDITIONAL RESOURCES: Current events articles in newspapers and magazines discuss ecological issues on an almost daily basis. Television shows often focus on the environment. These include episodes of

National Geographic, Nova, and *Innovation.* All of these may be purchased from different sources. If you cannot find them in catalogues, call The National Geographic Society in Washington, DC, as well as your local PBS station. Two specific recommendations are: The television show: "The Poisoning of America," (ABC News, 1988); and the book: Kupchella, Charles E. and Hyland, Margaret C., *Environmental Science*, Boston, Allyn and Bacon, 1989.

TEACHER'S GUIDE
CAMOUFLAGE: THE FIND THE DINNER GAME

GOAL: To comprehend how camouflage helps animals to survive.

STUDENT OBJECTIVES:
- To find camouflaged objects in a game of Find the Dinner.
- To discover that camouflage is an effective survival mechanism for many living things.

PRELAB DISCUSSION: In the fight for survival there are many animals that are far too small and weak to fight their likely predators in their food chain or web. The mechanism of camouflage is excellent both as a defense, by blending in, or as an offense, by pretending to be a more deadly plant or animal. In this activity, students, the consumers, will try to obtain as much food as possible in a given amount of time. Some of their intended prey will be camouflaged. The game is played best outdoors in a grassy setting, but can be adapted to any location by changing the colors and amount of time allotted. Take pictures!

GUIDE TO THE INVESTIGATION:

The class will need:

>   400 toothpicks: 100 each of four different colors, one of which is green

Scatter the toothpicks randomly on a large, grassy area that is not mowed closely. Mark off the habitat of the toothpicks. DO NOT TELL THE STUDENTS HOW MANY OF THE TOOTHPICKS THERE ARE OR THEIR COLORS. Give the students one minute to find as many toothpicks as they can. You can change the time if the size of the area and the number of students requires it. STUDENTS MAY NOT RUN OR PUSH!

Discuss the results. Hopefully, they will have caught many more of the prey that was not camouflaged. Be certain to have the students return to find the rest of the toothpicks.

VOCABULARY: camouflage, consumer, food chain, food web, habitat, predator, prey

ADDITIONAL RESOURCES: There are some wonderful books and films on camouflage. Suggestions include: " Hide and Seek," Oxford Scientific Films; Owen, Dennis Frank, *Camouflage and Mimicry*, Chicago, University of Chicago Press, 1980; and Ross, Wildas, *Can You Find The Animals?*, New York, Coward, McCann & Geoghegan, 1974.

220 Fundamentals of Science Activity Series

TEACHER'S GUIDE
THE BALANCE OF NATURE

GOAL: To understand that there is a balance of nature between predator and prey.

STUDENT OBJECTIVES:
- To learn that predators and prey are both important parts of food chains and food webs.
- To play the role of predators in a game, in order to determine the limitations to the size of a cougar population in a particular community.
- To determine a working definition for the term carrying capacity.

PRELAB DISCUSSION: Activities 10 and 11 are excellent as background for this exciting study of the relationship between cougars (predators), and their prey. Students should be familiar with the terms balance of nature, food chains, food webs, predator and prey. In addition, the term carrying capacity should be defined, briefly, before the activity, and in depth after the activity. This activity is best conducted outdoors, but may be played indoors if all of the desks are placed at the perimeter of the habitat. Monitor any legitimate "gentle pushing" if there is competition for the same prey. Be certain that it remains strictly under control, as students get very excited by this game. Have fun! Take pictures!

GUIDE TO THE INVESTIGATION:

Each class of 24-30 will need:

> 200 8 oz. paper cups (use fewer cups for smaller classes)
> black paper and string to make a blindfold
> marking pencil
> piece of loose-leaf size paper

Mark the paper cups on the bottom with the letter of the animal each represents (see the student procedure section). An excellent variation on this activity is to use different size cups as appropriate to the difficulty of catching a particular animal (size, camouflage and speed would be factors). Larger cups would be easier animals to spot and catch.

Review the rules of the game and Have the students complete the activity. Discuss the results.

VOCABULARY: balance of nature, carrying capacity, community, consumer, food chain, food web, habitat, predator, prey, producer

ADDITIONAL RESOURCES: The Discovery Channel has developed wonderful materials for its *Predators in Action* television show. Contact them at National Science Teachers Association conventions or call them.

TEACHER'S GUIDE
THE GROWTH OF PLANT ROOTS AND STEMS

GOAL: To understand some of the factors that influence the growth of plant roots and stems.

STUDENT OBJECTIVES:
- To determine that plant roots and stems are subject to the influence of gravity.
- To determine that plant stems are subject to the influence of light.
- To describe the role of unequal cell growth (due to auxins) in choosing the preferred direction of growth.

PRELAB DISCUSSION: In this activity the students will determine some of the factors which influence plant growth. Begin with a discussion of which way plant roots and stems grow. Ask the students why they grow that way. Depending on the level of the class, you may wish to discuss auxins, geotropism, hydrotropism and phototropism. This activity takes several days, but provides interesting results. Students should work in groups of two or more. Take pictures!

GUIDE TO THE INVESTIGATION:

Students will need:

petri dish	aluminum foil
16 corn seeds	scissors
flower pot	paper towels
3 small milk cartons	beaker or bowl
potting soil	water
scotch tape	marking pencil

Have the students complete the activity. Discuss the results.

VOCABULARY: tropism

ADDITIONAL RESOURCES: A fine discussion of plant hormones may be found in Arms, Karen and Camp, Pamela S., *Biology*, New York, Saunders College Publishing, 1987.

TEACHER'S GUIDE
YOU ARE THE SCIENTIST: THE GROW THE BEST PLANT CONTEST

GOAL: To understand what research is necessary to determine the optimum conditions necessary for seed germination and plant growth.

STUDENT OBJECTIVES:
- To design and carry out an original scientific experiment.
- To experimentally determine the optimum conditions for seed germination.

PRELAB DISCUSSION: This is an excellent long range activity for students to learn or review the elements of an experimental design. Review the scientific method. Focus on the importance of controls. In addition, elicit from the students what they know about what is necessary for seed germination and plant growth. Students, conducting this simple activity, have actually submitted their experimental designs in statewide competitions and have been awarded grants to carry out their research! Students may work alone, or in pairs or committees. Take pictures of the progress of all experiments every few days. A ruler next each plant adds to the pictures.

GUIDE TO THE INVESTIGATION:

Students should be encouraged to be creative when selecting a variable to test. Have students think of where and how plants grow all over the world, indoors and outdoors. The more variables the class tests, the more complete the conclusions will be. Some variables which may be tested include the type of soil or its pH, temperature conditions, moisture conditions, the amount, intensity and color of available light, the direction and depth of the seed, and external conditions such as the amount of oxygen (higher altitude plants usually have less) or pollution.

Discuss the possibilities with the class. Several days may be needed until students pick a variable which they want to test. Assist them with the experimental design. Include the type of data they will collect and a chart or other format in which to collect it. Experiments should test only one variable and should have a control group. Decide when the experiment will be ended, too. Approve all final plans and supervise the work.

Students will need seeds, such as those following, that germinate easily. Lima and kidney beans, which need to be soaked for a day before beginning, in order to loosen their seed coats. Seeds of grasses (garden, wheat, rye), which are small and need to be handled carefully. Also, seeds of radishes, marigolds, tobacco and peas. BE SURE TO USE A SINGLE VARIETY FOR EACH ACTIVITY. Pea seeds, for example, are often mixed tall and short, and so distort the results.

You will need to approve designs for which materials are available. Different growing mediums, pots (of different sizes and materials?), thermometers, lights and other resources may be needed.

VOCABULARY: conclusion, control, data, experimental design, germination, hypothesis, scientific method, variable

ADDITIONAL RESOURCES: The local garden store may be a source for unusual theories about how to grow plants. A more scientific approach may be found in: Babel, Nancy, *The Seed Starter's Handbook*, New York, Rodale Press, 1983.

Environmental chambers and their accessories permit highly controlled environments for experimentation. These may be purchased from Carolina Biological or other scientific supply companies. A local university may be willing to share their chambers either on loan or at their campus. It is worth the phone call to ask, and it is a wonderful experience for students to work at a campus. Be sure to arrange for appropriate supervision of the students.

TEACHER'S GUIDE
WHY SHOULD A NICE GREEN LEAF TURN RED OR YELLOW?

GOAL: To develop student understanding that the autumn colors of leaves are there all the time, but are usually masked by chlorophyll.

STUDENT OBJECTIVES:
- To extract the pigmentation from leaves.
- To use paper chromatography to separate the pigmentation colors.
- To identify the pigments found in chlorophyll.
- To realize that chromatography is a powerful scientific research tool.

PRELAB DISCUSSION: There are two areas to cover before students begin this activity. The first is a discussion of what colors leaves turn in the fall (in most parts of the country). Ask the students why the colors occur only in the fall. Without explaining more, tell them that they are going to investigate this. It is suggested that students look closely at the leaves of plants, such as coleus, which regularly display other colors. It is instructive, but not necessary, to look at such leaves under a microscope (wet mount slides) as well. The second area to cover is the procedures which students are to do. The teacher should prepare the pigment and solvent as a demonstration for the class.

BE SAFE! **Wear goggles and a laboratory apron. Do not inhale the alcohol or solvent fumes. Keep the room well ventilated. There should be no flames or high heat sources in the room.**

GUIDE TO THE INVESTIGATION:

The teacher will need:

defrosted package of frozen spinach	hot plate
1000 ml Pyrex beaker	water
250 ml Pyrex beaker	tongs
about 100 ml ethyl alcohol	goggles
petroleum ether	lab apron
acetone	forceps
250 ml (or more) container with cork	small funnel

The students will need:

large test tube with cork stopper	thumb tack
test tube rack	metric ruler
small capillary tube (or toothpick)	pencil
chromatography paper strips*	scissors
goggles	lab apron

(* heavy paper towel may be used, but is not as good)

PREPARING THE CONCENTRATED CHLOROPHYLL SOLUTION:

BE SAFE! **Alcohol is flammable. Never heat it on an open flame or directly on a hot plate. Use a "double boiler," a small beaker with alcohol inside a larger one containing water. Do not let the alcohol boil over the rim of its beaker. Ventilate the room. Do not breathe the fumes. Wear goggles and a lab apron.**

Place the spinach in a 250 ml beaker. Add just enough alcohol to cover the leaves. Place this beaker inside a 1000 ml beaker that is half-filled with water. Put the large beaker on a hot plate and boil the water, as shown in illustration 15-T. When the leaves have been bleached of their chlorophyll, carefully remove them from the beaker with forceps. Continue heating until the color of the alcohol is dark green. Use tongs to remove the beaker from the hot plate. Turn off the hot plate and allow the extract to cool.

PREPARING THE SOLVENT MIXTURE:

BE SAFE! **Petroleum ether is extremely volatile. Secure stoppers tightly. Do not inhale fumes. Ventilate the room. Wear goggles and a lab apron. Your students will need about 5 ml of this for each test tube (to a height of 1 cm). Do not make more than you need.**

BE SAFE! Mix 92 parts petroleum ether with 8 parts acetone. Place into a glass container. **Stopper it tightly.**

VOCABULARY: chlorophyll, chromatography, diffusion, pigment, solvent

ADDITIONAL RESOURCES: Information on chromatography can be found in Stryer, *Biochemistry*, W.H. Freeman & Co., San Francisco, 1981.

TEACHER'S GUIDE
IS BLACK INK REALLY BLACK?

GOAL: To develop student understanding that the color of black ink is the result of a mixture of substances.

STUDENT OBJECTIVES:
- To use paper chromatography to separate the colors of black ink.
- To realize that chromatography is a powerful scientific research tool.

PRELAB DISCUSSION:

This activity is a perfect sequel to the previous activity: "Why Should a Nice Green Leaf Turn Red or Yellow?"* Each shows the masking of colors. However, there is little external evidence that ink is a mixture. Thus, from this activity students comprehend the power of the paper chromatography technique. (* Note: It is not necessary to conduct the previous activity in order to do this one.) Tell the students that they are going to investigate whether or not black ink is really black. Use washable ink, as the experiment and cleanup are greatly simplified. Discuss the procedures which students are to follow.

BE SAFE! **Wear goggles and a laboratory apron.**

GUIDE TO THE INVESTIGATION:

The students will need:

large test tube (or graduated cylinder)	medicine dropper
test tube rack	metric ruler
black ink	pencil
chromatography paper strips*	scissors
small piece of aluminum foil	goggles
water	lab apron

* heavy paper towel may be used, but is not as good.

VOCABULARY: chromatography, diffusion, mixture, pigment, solution, solvent

ADDITIONAL RESOURCES: Information on chromatography can be found in Stryer, *Biochemistry*, W.H. Freeman & Co., San Francisco, 1981.

TEACHER'S GUIDE
MESSAGES IN INVISIBLE INK

GOAL: To understand the chemistry of invisible ink.

STUDENT OBJECTIVES:
- To prepare invisible inks.
- To write and retrieve messages using invisible inks.
- To learn the chemistry of invisible inks.

PRELAB DISCUSSION: This is an excellent follow up to Activity 16, *Is Black Ink Really Black*. Disappearing inks rely on several mechanisms. Many are indicators which are insoluble in water. When dissolved, usually in alcohol, they dry to a clear or white powder. They will turn color in the presence of weak acids or bases. Other inks are affected by heat. In this activity you may choose one or all of the parts to do.

BE SAFE! **Part 5 should be done only as a teacher demonstration.** Practice each part before you or the students do it. Students should work in pairs for the activity. **Goggles should be worn.**

GUIDE TO THE INVESTIGATION:

Students will need: For all parts...

loose leaf paper
2 250 ml Pyrex beakers
fountain pen (not ball point)
test tube rack
goggles

cotton balls and swabs
scoopula or spoon
two test tubes
lab apron

In addition, they will need:

Part 1:

coffee filter
sal soda*
methyl or ethyl alcohol

water
phenolphthalein

* also called washing soda: $Na_2CO_3 \cdot 10 H_2O$

Part 2:

coffee filter
balance
methyl or ethyl alcohol

water
sodium hydroxide
thymolphthalein

Part 3:

cobalt chloride
hot plate

water

Part 4:

copper sulfate
ammonium hydroxide

water

Part 5:*

potassium nitrate
Bunsen burner or hot plate

water

BE SAFE! * teacher demonstration ONLY

Complete any or all parts. Discuss the results.

BE SAFE! Instructions for **Part 5: Wear goggles (Teacher demonstration only)**

1. Dissolve a small amount of potassium nitrate in a little water in a test tube.

2. Dip the pen point into the solution and write a message on loose leaf paper. Dip the pen after each letter.

3. Let the paper dry.

BE SAFE! 4. Heat the paper to the point where it starts to scorch. **Do not burn the paper. Keep water at hand in case you set the paper on fire! Do not keep flammables near the paper.**

5. Show the class what happens.

VOCABULARY: acid, base, indicator

ADDITIONAL RESOURCES: The chemistry of acids, bases and indicators may be found in any high school chemistry text, such as Dorin, Henry, *Chemistry, The Study of Matter*, Allyn and Bacon, 1989.

TEACHER'S GUIDE
MODELS OF ATOMS

GOAL: To understand the purpose of creating models of atoms.

STUDENT OBJECTIVES:
- To describe the reasons for using models.
- To create models of the first 20 atoms of the periodic table.
- To predict the formulas for simple molecules by using their models.

PRELAB DISCUSSION: This activity permits students to see at first hand the reason for the formulas of simple molecules. In addition, it increases their comprehension of the need for good models. They should be familiar with the terms and locations of the nucleus, neutrons, protons, and electrons. They should know the term atomic number, and that it tells the number of protons or electrons in an atom. Students should work in groups of two to four. Permit them to talk quietly, in order to decide on the next step in their model construction. Each group will need about 2 m by 2 m of floor space. Thin bamboo strips may be purchased in hobby and other stores. Styrofoam balls may be purchased in bulk from manufacturers and then spray painted.

GUIDE TO THE INVESTIGATION:

Students will need:

bamboo rings (cardboard may be used instead) with these diameters: 5 K rings at 8 cm; 5 L rings at 15 cm; 2 M rings at 24 cm; and 2 N rings at 30 cm.

about 60 Ping–Pong or Styrofoam balls (flat cardboard can be used): Color is more visually effective than white. If possible, use 20 of one color, 20 of another, and 10 each of two others.

periodic table, or list of elements 1-20 with their atomic numbers, but WITHOUT ELECTRON CONFIGURATIONS.

paper

scissors

marking pen

Have the students follow the procedures. Discuss the results.

VOCABULARY: atom, atomic number, bond, electron, model, proton, neutron, nucleus

ADDITIONAL RESOURCES: The history of atomic theory, as well as the electronic configuration for each atom may be found in books such as: Dorin, Henry, *Chemistry, The Study of Matter*, Allyn and Bacon, 1989. The electron structure may be found on many periodic charts as well, but should not be given to the students before this activity is completed.

TEACHER'S GUIDE
RACE CAR TRIALS

GOAL: To have students comprehend the concept of velocity.

STUDENT OBJECTIVES:
- To recognize that average velocity is the total distance traveled (change in position) divided by the time it took to go that distance.
- To determine the average velocity of toy racing cars.
- To understand the difference between average and instantaneous velocity.

PRELAB DISCUSSION: This experiment allows students to relate science to concepts encountered daily, such as velocity and friction. It is easy to comprehend and interesting to the students. It will take several periods to do the entire experiment. It can be done in parts, or each part can stand alone. It may also be expanded in many ways. For example, the activity may be extended in order to develop the concept of acceleration. Before the activity, explain the difference between speed (distance divided by time) and velocity (speed with a direction, such as two meters per second forward). Students may need assistance when producing their graphs. For this experiment, they should work in groups of two or more.

GUIDE TO THE INVESTIGATION:

Students will need:

>   Matchbox car
>   three meters of race car track
>   or stiff cardboard
>   plastic wrap (for friction trials)
>   sandpaper (for friction trials)
>   pennies
>   books to elevate the track

>   tape (double stick, if
>   available)
>   stopwatch
>   meter stick
>   graph paper

Students should complete the procedure section of the activity.

Have students graph their data and record their conclusions. Discuss their results.

VOCABULARY: average velocity, friction, instantaneous velocity, velocity

ADDITIONAL RESOURCES: Several interesting motion experiments may be found in *Conceptual Physics Laboratory Manual* (Paul Robinson, Addison-Wesley, 1987).

TEACHER'S GUIDE
LET'S MOVE THE PIANO: THE INCLINED PLANE

GOAL: To have students recognize the advantage of using a simple machine: the inclined plane.

STUDENT OBJECTIVES:
- To discover the relationship between the height of an inclined plane and its effectiveness as a machine.
- To discover the relationship between the length of an inclined plane and its effectiveness as a machine.

PRELAB DISCUSSION: In this experiment, the first of three involving simple machines, students will measure the ability of different inclined planes to help move loads. Then they will graph their results in order to obtain a strong sense of how the effectiveness of a ramp is related to its height and its length.

Before conducting this experiment, review relevant graphing techniques, as well as the concepts of work (force x distance), effort and load.

GUIDE TO THE INVESTIGATION:
Students will need:

laboratory balance	books for height
model train flatcar (or substitute car that can hold weights)	meter stick
	ruler
two pieces of flat board (30 and 60 cm long)	graph paper
spring scale	
50 gram and a 100 gram mass	

Have the students complete the activity. Discuss the data and the graphs. You may wish to discuss or demonstrate the incorporation of inclined planes into more complex machines.

VOCABULARY: effort, inclined plane, lever, load, pulley

ADDITIONAL RESOURCES: Several books discuss the history of machines. Recently, the connections between inventions and subsequent developments in human history were explored wonderfully in television shows such as the P.B.S. series, *The Ascent of Man*.

TEACHER'S GUIDE
LET'S MOVE THE PIANO: THE LEVER

GOAL: To have students recognize the advantages of using a simple machine: the lever.

STUDENT OBJECTIVES:
- To comprehend the different classes of levers.
- To learn how to develop the most effective use from the human arm, a typical class 3 lever.

PRELAB DISCUSSION: In this experiment, the second of three involving simple machines, students will measure the forces needed at different locations on the human arm to lift a load. The data will be graphed in order to determine the optimum use of this lever. Begin with a review of the concepts of work (force times distance), effort, load and relevant graphing techniques. It is suggested that this activity follow activity 20, which investigates the inclined plane.

Discuss the classes of levers (presented on the student project sheet) and some examples of each. Note that for this experiment, students should work in groups of two or more.

GUIDE TO THE INVESTIGATION:

Students will need:

500 g mass	meter stick
spring scale with a capacity of 20 newtons	ruler
	graph paper
narrow watchband (or an equivalent round loop to slide up and down the forearm)	

Have the students complete the activity. Discuss the data and graphs. You may wish to discuss or demonstrate the incorporation of levers into more complex machines.

VOCABULARY: effort, fulcrum, inclined plane, lever, load, pulley, work

ADDITIONAL RESOURCES: Several books discuss the history of machines. Recently, the connections between inventions and subsequent developments in human history were explored wonderfully in television shows such as the P.B.S. series, *The Ascent of Man*.

TEACHER'S GUIDE
PULLEY TUG OF WAR

GOAL: To have students recognize the advantages of using a simple machine: the pulley.

STUDENT OBJECTIVES:
- To understand the use of simple and complex pulley arrangements.
- To discover the relationship between the number of effective rope strands on a pulley system and the usefulness of the system.
- To apply their knowledge to building a system for a pulley tug of war.

PRELAB DISCUSSION: In this experiment, the last of three involving simple machines, students will determine the effectiveness of various pulley systems. They will then construct a pulley system for their tug of war. Each part of this activity is independent of the other. It is suggested that the parts be done in order. If only the tug of war is done, students should be told about the relationship between the number of effective strands of rope and the usefulness of the entire system. Students should review the concepts of work (force x distance), effort, and load. Students should work in groups of two or more for Part 1. Part 2 should be a class activity.

GUIDE TO THE INVESTIGATION:

Students will need: (**Part 1**)

ringstand and pole	two single pulleys
utility clamp	wire
one kilogram mass	scissors
spring scale with a 10 newton capacity	books for supportive weight
	fish line (or strong cord)

Students will need: (**Part 2**)

long, heavy rope
heavy object to move
large spools for pulleys (as electrical wire spools)

Have the students complete the activity. Discuss the data and conclusions for Part 1, and the results of Part 2. You may wish to have students investigate the incorporation of pulleys into more complex machines.

VOCABULARY: effective strand, effort, inclined plane, lever, load, pulley, work

ADDITIONAL RESOURCES: Science museums, such as The Franklin Institute in Philadelphia and the Exploratorium in San Francisco, often have imaginative, interactive machines for direct investigation. These museums are very responsive to visits and requests for assistance in setting up similar demonstrations.

As mentioned previously, the connections between inventions and subsequent developments in human history have been explored in books and in television shows such as the P.B.S. series, *The Ascent of Man*.

TEACHER'S GUIDE:
INTERESTING INERTIA

GOAL: To develop student knowledge of inertia.

STUDENT OBJECTIVES:
- To explain the concept of inertia.
- To develop inertia based explanations for everyday events as well as for several "magic" tricks.

PRELAB DISCUSSION: Inertia experiments are intrinsically motivating. Begin by asking students: What would be easier to push from a speed of zero to a speed of 10 km / hr, a toy truck or a bulldozer? Continue with: Which would you rather be assigned to deflect from its straight line path, a toy truck or a bulldozer, if each goes at 10 km / hr? Define inertia and present Newton's first law of motion: An object in motion will remain in straight line motion and an object at rest will remain at rest unless acted on by a (net external) force.

BE SAFE! Students should be asked to conduct the investigations presented, **but they should take extreme care that neither they nor others are in the path of moving objects.** After the students are done, it is often interesting to ask for ideas for additional inertia experiments.

GUIDE TO THE INVESTIGATION:

Students will need:

- string
- inertia ball (or a large mass with secure hooks on the top and bottom)
- strong support
- wooden blocks
- metal cup
- mallet with two hitting sides to its head (a hammer will do but is less effective)
- tomato or egg (optional)
- pile of magazines
- piece of wood about 8 to 12 cm on a side
- nail
- stack of books
- pile of sponges or foam rubber
- metric ruler
- some water

Arrange for a safe area to conduct the experiments. Review safety precautions with the students, then have them carry out the procedures.

Discuss the student findings in terms of inertia. Ask students how the "magic" trick of pulling a table cloth out from under the table settings works. You may want to try this as a demonstration. Use old dishes!

VOCABULARY: inertia

ADDITIONAL RESOURCES: Sir Isaac Newton's *Philosophiae Naturalis Principia Mathematica* (London, 1687) has an outstanding explanation of the concept of inertial mass. The book is fascinating in many ways, including the use of geometry to prove to others that the calculus, which he invented to prove his theories, yielded the same answers as would the far more time consuming geometrical proofs.

TEACHER'S GUIDE
ACTIONS AND REACTIONS

GOAL: To develop student understanding of Newton's third law of motion.

STUDENT OBJECTIVES:

To demonstrate that actions have equal and opposite reactions.

To describe how this law can be used to propel rockets.

PRELAB DISCUSSION: This activity is a great deal of fun for the students. Ask students to describe an action, such as making a face at someone, and a reaction. Define force, and indicate that actions in science are forces.

BE SAFE! **Make certain that students wear goggles and a lab apron. Follow all precautions for the vinegar—baking soda reaction: The stopper must fit into the bottle, but should NOT be pushed in tightly. Students should stand clear of the cork and the bottle once the reaction starts.** Students should work in pairs.

GUIDE TO THE INVESTIGATION:

Students will need:

2 spring scales	vinegar
4 plastic straws	baking soda
balloon	graduated cylinder
thin string	narrow mouth bottle
tape	stopper for the bottle
scissors	glycerol or petroleum jelly
piece of filter paper (or tissue)	goggles
balance	lab apron

BE SAFE! Obtain bottles and then find the amount of vinegar and baking soda needed by trying the activity in advance. A half liter bottle is a good size. A good starting point is 1 ml of vinegar to each gram of baking soda. When the bottle is on its side, the vinegar should be two to three centimeters deep. **The vinegar should not come up as high as the neck, otherwise the gas pressure is on the sides and not on the stopper!!**

Have the students conduct the activity. Review the results. You may wish to discuss the reaction of vinegar (dilute acetic acid) and baking soda (sodium bicarbonate). The products are sodium acetate, water and carbon dioxide.

VOCABULARY: action, force, reaction

ADDITIONAL RESOURCES: More advanced action—reaction demonstrations are found in high school physics books. Caution should be used when conducting these. For example, the balloon on a string can be done as a carbon dioxide canister on a wire, but the compressed gas requires awareness of its potential hazards.

It is interesting to read Newton's original thoughts on action and reaction. These may be found in *Philosophiae Naturalis Principia Mathematica*, London, 1687.

TEACHER'S GUIDE
GETTING A LIFT FROM AIR PRESSURE

GOAL: To understand that lift is a result of differences in air pressure.

STUDENT OBJECTIVES:

To create and observe the results of pressure differences on objects that are denser than air.

To explain lift using Bernoulli's principle.

PRELAB DISCUSSION: This activity involves several parts, all of which are fun for the students, in as much as the results are generally unexpected. Define pressure as force per area and review the concept of density. It is best to discuss Bernoulli's principle, that pressure decreases as the speed of a fluid increases, after the students make their observations, as this preserves the impact of the results. Students should work in groups of two or more. Take pictures!

GUIDE TO THE INVESTIGATION:

Students will need:

piece of paper, about 10 x 15 cm	2 Ping-Pong balls
index card, any size	ring stand and clamp
2 plastic straws	wooden or metal rod
beaker	metric ruler
water	tape
food coloring	scissors
funnel	string

Have the students conduct the activity. Review the results.

VOCABULARY: Bernoulli's principle, density, lift, pressure

ADDITIONAL RESOURCES: Bernoulli's principle is a bit more complex than the statement above. A detailed analysis can be found in college physics texts such as: Morgan, Joseph, *Physics*, Boston, Allyn and Bacon. Application of Bernoulli's theorem are especially important in the design of airplanes. A visit to a wind tunnel, or a film showing their testing procedures enhances this topic.

General Science Basic Activities 247

TEACHER'S GUIDE
COLLAPSING CANS

GOAL: To demonstrate how a difference in pressure results in a force.

STUDENT OBJECTIVES:
- To describe the difference between force and pressure.
- To create a pressure difference great enough to collapse a can.
- To calculate the external pressure on a can.

PRELAB DISCUSSION: This experiment is often done as a demonstration, but can be done by individuals. Both methods will be discussed here. Begin by describing the difference between force and pressure (force per unit area). Tell the students that the pressure inside the can in this experiment is to be reduced by driving out many air molecules with heat.

BE SAFE! **It is important to keep water in the bottom of the can so that it does not cause a hazard. Make certain the water does not evaporate totally. Never heat a closed can. Do not attempt to re-form the can by heating. Safety goggles are necessary. Adequate gloves, tongs or potholder mitts should be used to handle the can.**

GUIDE TO THE INVESTIGATION:

METHOD 1—DEMONSTRATION

BE SAFE! 1. Thoroughly rinse a rectangular gallon can, such as a ditto fluid can, with water. **Be certain that no vapors are left in the can.**

2. While heating the can, start a stream of cold water so that the can may be placed under the stream as quickly as may safely be accomplished.

GUIDE TO THE INVESTIGATION:

METHOD 2—STUDENT ACTIVITY

1. Students will need a rinsed empty aluminum soda can, cold water, a source of heat, a ruler, goggles and mitts (or equivalent, to hold a hot can).

 2. Emphasize the safety aspects of the activity. If students do not have a sink at their seat have them use a bucket or beaker of cold water placed at their desk. The hot can should **NOT** be fully submerged. **Be certain students do not touch the hot can.**

VOCABULARY: force, pressure

ADDITIONAL RESOURCES: *The Flying Circus of Physics* (Jearl Walker, John Wiley & Sons) has many fascinating questions about pressure, among other topics. If you are as curious as most of us, obtain the version with answers.

TEACHER'S GUIDE
THE STATIC ELECTRIC MAGIC WAND

GOAL: To increase student understanding of the interaction of charged and neutral objects.

STUDENT OBJECTIVES:
- To produce a static electric charge on objects.
- To observe and explain the interactions of objects with a static electric charge and neutral objects.

PRELAB DISCUSSION: This activity is wonderful fun, and leads to the exploration of numerous phenomena (see ADDITIONAL RESOURCES). It is suggested that students receive no background knowledge of static electricity before conducting the investigation, but rather, that the observed behavior be explained after the conclusion of the activity. Thus, the investigation could be used as the starting point for a unit on static electrical effects. or on electricity in general.

GUIDE TO THE INVESTIGATION:

Students will need:

> balloon
> 2 x 4 piece of wood (a meter stick works well, too)
> small block of wood to support the long piece of wood (perhaps 5 x 2.5 x 2.5 cm)
> rubber rod and a piece of fur (or glass rod with silk)
> sink

This activity works best on days with low humidity, and if possible, should be done on such a day.

Have the students conduct the investigation.

Carefully explain the students' observations.

As a supplement, an exact comparison of the relative strength of static electrical versus gravitational attraction, in terms of orders of magnitude, is informative. It takes the gravity of the entire planet to hold a person down, but a small static charge on a balloon is enough to "defy" gravity and have it stick to a wall.

VOCABULARY: charge, conductor, insulator, neutral, static electricity

ADDITIONAL RESOURCES: Many static electric and discharge phenomena are noted and explained in: Walker, Jearl, *The Flying Circus Of Physics WITH ANSWERS*, New York. John Wiley & Sons, 1977.

TEACHER'S GUIDE
THE BUBBLE TELEGRAPH

GOAL: To understand the historical development of the telegraph.

STUDENT OBJECTIVES:
- To make a bubble telegraph.
- To describe the development of the telegraph.

PRELAB DISCUSSION: The historical development of the telegraph is fascinating. Most people know that the modern telegraph was invented in 1844 by Samuel F. B. Morse, but few know that the idea of using electricity to send messages is far older. Review this development with the students. In 1878, in Italy, Alessandro Volta invented the voltaic cell, or simple battery. He noted soon after that wires from this cell to a conducting liquid produced bubbles, seemingly independent of the distance. He believed that these bubbles could be used to send messages. In 1809, Samuel T. von Soemmering, a German, used thirty-five separate sets of wires, one for each letter and digit, to send bubble messages over a 300 meter distance. This impressed the world. The thought of using one set of wires and a code had not occurred yet. For this activity, in the interest of developing student manipulative skills, teach the students how to use a wire cutter safely. Students should work in groups of two or more for this activity.

BE SAFE! **Be certain to review electrical safety procedures. Goggles should be worn.**

GUIDE TO THE INVESTIGATION:

Students will need:

beaker	switch
insulated copper wire	water
(e.g., bell wire)	goggles
table salt	
9 volt battery	

Have the students conduct the procedures. Review the results.

This activity is an excellent companion to the next one, which is about batteries.

VOCABULARY: battery, conductor, current

ADDITIONAL RESOURCES: A discussion of current events in communications is appropriate with this activity. A visit to a telephone, satellite communication or fax facility is in order. It may be that the telegraph today has become obsolete as quickly as the Pony Express did when the telegraph was invented.

TEACHER'S GUIDE
THE LEMON BATTERY

GOAL: To understand the historical development of the battery.

STUDENT OBJECTIVES:
- To make a voltaic cell
- To describe the development of the battery.

PRELAB DISCUSSION: The historical development of the use of electricity is highly interesting. Most people know about the (rather stupid) work of Ben Franklin and his kite flying in a thunderstorm, but fewer know of the observations of Galvani and Volta. In light of their work, stress to the students the importance of being good observers. In the interest of developing student manipulative skills, teach the students how to use a wire cutter safely. In addition, teach them how to connect electrical meters. Students should work in pairs for this activity.

BE SAFE! **Be certain to review electrical safety procedures. Goggles should be worn.**

GUIDE TO THE INVESTIGATION:

Students will need:

> 300 ml concentrated lemon juice
> 400 ml beaker
> copper strip
> zinc strip
> insulated copper wire (e.g., bell wire)
> galvanometer (or 1.1 volt electric clock with leads)
> lemon (optional: orange, potato, soda, etc.)
> goggles
> (optional: sensitive voltmeter, other metal strips)

Have the students conduct the procedures. Review the results.

This activity is an excellent companion to the previous activity about the telegraph.

VOCABULARY: conductor, current, electrolyte

ADDITIONAL RESOURCES: Be certain to review electrical safety procedures. Goggles should be worn. The commercially available and popular Potato Clock can be used to conduct this activity. It is used more effectively as an example of how very simple and well-known science, at a level understandable by virtually all students, can be turned into a successful product by using a little creativity.

To determine the voltage between any two metals, find the difference in the electrochemical series in any chemistry text.

TEACHER'S GUIDE
CONDUCTING COMPOUNDS

GOAL: To understand what kinds of materials are good conductors and what kinds of materials are not good conductors.

STUDENT OBJECTIVES:
- To find some materials which are good conductors.
- To classify the good conductors by type of material.

PRELAB DISCUSSION: Materials are categorized as good conductors, semi-conductors and poor conductors, or insulators. (Most materials can carry electricity, given a high enough electrical energy level, or voltage, so the term poor conductor is preferable to non-conductor.) This activity will allow the students to determine that metals and dissolved salts are generally good conductors. In the belief that students should learn to manipulate electrical circuits, it is recommended that they work in groups of two or more and not use ready-made boards. The students should be shown how to strip a wire safely and how to connect wires to batteries.

BE SAFE! **Students should be told not to hold bare ends of wires that will complete a circuit and not to put their hand in a liquid that might complete a circuit or have a "hot" wire in it.** A voltmeter or ammeter may be used in place of a light bulb. If meters are used, students must be shown how to hook them up. Other chemicals may be used, but greater precautions for the students may be needed. **Goggles should be worn.**

GUIDE TO THE INVESTIGATION:

Students will need:

> 2 1.5 volt dry cell batteries
> battery holder(s) if needed
> insulated copper wire
>   (e.g., bell wire)
> wire cutters
> 3 volt miniature light bulb
> miniature base
> small screwdriver for base
> beaker

> aluminum foil
> iron nail
> penny
> piece of plastic
> piece of glass
> wool
> cotton
> piece of ceramic

water (distilled is best)
salts (sodium chloride, and others)
other materials to test, as desired
sugar
pieces of paper
goggles

Have the students conduct the investigation.

Review the results. This activity can be used as a bridge to the concepts of ions and of electroplating.

VOCABULARY: battery, circuit, conductor, insulator, organic material, salt

ADDITIONAL RESOURCES: Utilities generally provide speakers and exciting demonstrations for assembly programs on electrical safety. They are sources of free pamphlets on all aspects of electricity, as well. If your utility does not have these items contact Edison Electric Institute, in Washington, DC.

TEACHER'S GUIDE
SIMPLE SERIES CIRCUITS

GOAL: To develop student understanding of the behavior of batteries and resistances in series circuits.

STUDENT OBJECTIVES:
- To establish a series circuit.
- To measure the voltage of dry cells connected in series.
- To determine the voltages and currents in a series arrangement of resistances.

PRELAB DISCUSSION: This experiment is designed to introduce students to simple series circuits. In the belief that students should learn to manipulate electrical circuits, it is recommended that they work in groups of two or more and not use ready-made boards. Thus, students should be shown how to strip a wire safely. In addition, they will need to know how to connect and read voltmeters and ammeters. The range of the meters will depend on your available resistors. Conversely, if you are limited by your meters, use resistors or rheostats to adjust the experimental parameters.

GUIDE TO THE INVESTIGATION:

Part 1:

Students will need:

> three 6-volt dry cell batteries DC voltmeter
> insulated copper wire wire cutters
>   (e.g., bell wire)
> simple switch (e.g., SPST knife switch)

Part 2:

Students will need:

> the materials in Part 1
> DC ammeter
> three resistors (it is recommended that miniature light bulbs and
>   bases be used in order to reinforce the results visually. If this is

done, then a screwdriver with a blade appropriate for the bases will be needed as well).

Have the students complete the activity.

Review the results while reinforcing the major advantage (greater potential difference), and major disadvantage (short life) of connecting cells in series.

VOCABULARY: ammeter, battery, dry cell, electrode, electrolyte, series circuit, resistor, voltmeter, wet cell

ADDITIONAL RESOURCES: Many areas of the country have companies which manufacture printed circuits in various sizes. An exciting field trip can be arranged to investigate the parts and part functions of such circuits.

TEACHER'S GUIDE
FUN WITH MAGNETS

GOAL: To develop student knowledge of the law of magnets and magnetic fields.

STUDENT OBJECTIVES:
- To investigate and explain the interactive effects of magnets.
- To describe the magnetic field surrounding a single bar magnet and the fields surrounding various interacting bar magnets.

PRELAB DISCUSSION: This is another laboratory in which the students have a great deal of fun while they are learning. Motivational demonstrations are easy to conduct, and are fascinating. For example, if you have a very powerful magnet at hand, you might pull around a filing cabinet or suspend yourself a few inches off the ground by holding the magnet and bringing it to a **SECURELY** fastened overhead iron rail (Make certain that you do not have any heavy object fall on you. Do not permit students to do this!). In any event, be sure to demonstrate the "suspended magnets", shown in illustration 32-T, without explaining what the students observe: take five circular magnets with small holes in the center (available at minimal cost from Radio Shack or other electronic supply stores) and place them on a pencil with opposite poles facing each other. They, of course, appear to float, and interact beautifully if they are pushed.

GUIDE TO THE INVESTIGATION:

Part 1:
Students will need:

> two bar magnets (as strong as possible, and marked N-S)
> ringstand, pole and clamp
> wooden rod (15-30 cm) to place in the clamp
> string

Part 2:
Students will need:

two bar magnets
iron filings
some plain white paper

Have the students complete the investigation.

Discuss the results. Highlight the way magnets exhibit "action at a distance." Explain the inverse square relationship of the field strength with distance, and that the strength never reaches zero.

VOCABULARY: magnetic field, pole

ADDITIONAL RESOURCES: In this topic, current events articles in newspapers and journals are omnipresent. From the superconducting magnets which levitate new trains (with different design concepts) in Asia and Europe, to the magnetic effects of the earth's core on geological studies; and from the "science fiction like" Meissner effect, to the use of magnetic "bottles" in fusion research, magnetism is pushing the frontiers of science. Some ideas to explore are in the *SUGGESTIONS FOR FURTHER STUDY* section, but many more can be investigated.

TEACHER'S GUIDE
MAGNETIC MATERIALS

GOAL: To develop student knowledge of which materials are magnetic.

STUDENT OBJECTIVES:
- To identify common magnetic materials.
- To find magnetic material in an unexpected place: a food product.

PRELAB DISCUSSION: This is yet another laboratory in which the students have a great deal of fun and some surprises while they are learning. It is helpful. but not necessary, to student comprehension of this activity if they have completed the *Fun With Magnets* activity. Again, motivational demonstrations are easy to conduct, and are fascinating. For example (as stated in the previous activity), if you have a very powerful magnet at hand, you might pull around a filing cabinet or suspend yourself a few inches off the ground by holding the magnet and bringing it to a SECURELY fastened overhead iron rail (Make certain that you do not have any heavy object fall on you. Do not permit students to do this!). Enjoy this lab. Take pictures!

GUIDE TO THE INVESTIGATION:

Part 1:

Students will need:

> bar magnet (as strong as possible, and marked N-S)
> ringstand, pole and clamp
> wooden rod (15-30 cm) to place in the clamp
> thin string ("invisible" thread or thin fishline enhances the activity)
> scissors
> book or other weight to hold down the string
> paper clip
> piece of aluminum foil
> thin glass cover plate (or any thin sheet of glass)
> thin sheet of plastic
> thin sheet of steel (the top of a can works, but be careful of sharp
>   edges)
> thin sheet of copper (use a penny if no other source is available)

piece of fabric
piece of paper and any other material which you wish to test

Part 2:

Students will need:

> beaker, 400 ml or larger (or a plastic or glass bowl)
> 250 ml of any iron-fortified cereal (100 % RDA for iron)
> bar magnet
> stirrer (glass or wood only)
> string
> water

Have the students complete the investigation. Discuss the results. Highlight the way magnets exhibit "action at a distance." Explain the inverse square relationship of the field strength with distance, and that the strength never reaches zero.

VOCABULARY: magnetic field

ADDITIONAL RESOURCES: In this topic, current events articles in newspapers and journals are omnipresent. From the superconducting magnets which levitate new trains (with different design concepts) in Asia and Europe, to the magnetic effects of the earth's core on geological studies; and from the "science fiction like" Meissner effect, to the use of magnetic "bottles" in fusion research, magnetism is pushing the frontiers of science. Some ideas to explore are in the *SUGGESTIONS FOR FURTHER STUDY* section, but many more can be investigated.

TEACHER'S GUIDE
MAKING MAGNETS STRONGER

GOAL: To understand some of the factors which affect the strength of electromagnets.

STUDENT OBJECTIVES:
- To determine the effect of different cores on the strength of electromagnets.
- To determine the effect of the number of turns of wire on the strength of electromagnets.

PRELAB DISCUSSION: The strength of electromagnets may be increased by using a core of iron or steel (an alloy of iron and nickel). It may be increased by increasing the number of coils of wire and by increasing the current. In this activity, students will investigate the core and coil factors. The activity is extended easily to investigate current, if the materials are available. In the belief that students should learn to manipulate electrical circuits, it is recommended that they work in groups of two or more. The students should be shown how to strip a wire safely and how to connect wires to batteries. Graphing procedures should be reviewed.

BE SAFE! **Students should be told not to hold the bare wires and to be aware that the coils and wire may get very hot.**

GUIDE TO THE INVESTIGATION:

Students will need:

1.5 volt dry cell battery	aluminum foil
battery holder, if needed	paper
bare copper wire	cardboard tube
(e.g., stripped bell wire)	wooden dowel
wire cutters	glass rod
plastic rod	graph paper
10 cm long thin iron or steel nail	
10 cm long thick bolt of the same material	
identical steel paper clips	
other materials to test, as desired	

Have the students conduct the investigation. Review the results.

VOCABULARY: core, electromagnet

ADDITIONAL RESOURCES: It is fascinating to see science applied. Electromagnets are used everywhere, and students can enjoy visiting giant versions or taking smaller ones apart. Candidates include old telephones, telegraphs, motors, and door bells. Be certain to supervise these activities, as many electrical, mechanical or chemical parts can be dangerous.

TEACHER'S GUIDE
THE SPEED OF HEAT TRANSFER

GOAL: To have students comprehend that the rate of heat transfer depends on the temperature difference between objects.

STUDENT OBJECTIVES:
- To observe and describe the methods of heat transfer.
- To determine that the rate of heat transfer depends on the temperature difference between objects.

PRELAB DISCUSSION: In this activity, the teacher begins by defining and demonstrating the three methods of heat transfer: conduction, convection and radiation. Many discussions can be used to stress the importance of heat transfer to our lives. For example, without radiation, the earth could not be warmed by the sun. The student part of this activity is a modified diffusion activity used to verify that the transfer of heat is fastest when the temperature difference is greatest. Students should work in groups of two or more for this activity.

BE SAFE! **Students should wear goggles and be cautioned about safety procedures when working with hot plates and hot water.**

GUIDE TO THE INVESTIGATION:

The teacher will need:

goggles	2 juice cans
ring stand and clamp	2 thermometers
metal rod	white and black paper
candle wax	scissors
knife	tape
Bunsen burner	150 watt light and base
convection tube (or similar device)	food coloring

The students will need:

hot plate	clock or timer
Pyrex beakers, 100 ml and 1 liter	food coloring
dropper bottle with dropper	water
forceps	scissors
string	goggles

BE SAFE! The teacher should begin by demonstrating heat transfer. **Wear goggles.** To demonstrate conduction, cut four small, equal pieces of wax and space them equally on a metal rod at 1/4, 1/2, 3/4 and all the way down the rod. As shown in illustration 35-1, clamp the rod horizontally on a ring stand. Heat the end without the wax. Have the students record their observations.

For convection, fill the convection tube with water. Clamp it as shown in illustration 35-2. Add a drop of food coloring, and heat the tube. Have the students record their observations.

For radiation, cut and tape black paper tightly around the sides and ends of a can. Use white paper for the other can. Insert a thermometer into each. As shown in illustration 35-3, suspend a light so that it shines equally on both cans. Call out the temperature of each can every minute or so. Have the students record the data.

Shut the light when the difference in absorption of radiant energy is apparent, but in any case, before the thermometers near their capacity!

Have the students carry out their procedures. Review the results.

VOCABULARY: conduction, convection, diffusion, heat, radiation

ADDITIONAL RESOURCES: Scientific supply houses have many transfer of heat kits, including a conduction by different metals apparatus, convection tubes, and black-and-shiny cans or thermos bottle arrangements for showing radiation.

TEACHER'S GUIDE
HEAT AND EXPANSION

GOAL: To have students understand that heat causes objects to expand.

STUDENT OBJECTIVES:
- To use heat to expand solids, liquids and gases.
- To describe the different rates of expansion of solids, liquids and gases.

PRELAB DISCUSSION: Begin by asking the students if they have ever noticed that railroads, bridges, highways and sidewalks have expansion joints. These are to allow for expansion during hot weather. If they are missing, or poorly designed, the metal or concrete will buckle. In this activity, students will compare the expansion rates of solids, liquids and gases. There are two common methods of displaying the expansion of solids; a thermal expansion of solids apparatus and a ball-and-ring apparatus. The latter is far simpler to use and to understand, and is inexpensive. For the expansion of liquids and solids, place the tubing into the stoppers in advance. This is a relatively risky procedure for students to do. Use glycerol as a lubricant and **take all proper safety precautions.** The glass tubing should protrude about 3 cm from the bottom of each stopper. Students should work in groups of two or more for this activity.

BE SAFE! **Students should wear goggles and be cautioned about safety procedures when working with hot plates and hot water, and about the heating of alcohol.**

GUIDE TO THE INVESTIGATION:

Students will need:

> ball-and-ring apparatus alcohol
> hot plate water
> Pyrex beaker, 400 ml or larger glycerol
> 3 Pyrex test tubes thermometer
> ring stand, pole and 2 clamps slotted cork
> rubber band food coloring
> marking pencil goggles

3 20 cm long pieces of glass tubing for the stoppers
1-hole rubber stoppers for the tubes

Have the students carry out their procedures. Review the results.

VOCABULARY: expansion, heat

ADDITIONAL RESOURCES: Scientific supply houses carry the ball-and-ring apparatus. A simple version of a thermal expansion apparatus is the *Introductory Physical Science* version, also available through supply houses. A lecture from an engineer who builds railways, bridges or roads is an interesting follow-up to this activity.

TEACHER'S GUIDE
RECYCLING PAPER

GOAL: To understand the value of recycling to the planetary ecosystem.

STUDENT OBJECTIVES:
- To make new paper from old.
- To explain how recycling saves energy, and resources while reducing pollution.

PRELAB DISCUSSION: This relatively simple activity demonstrates the ease with which vast good can be accomplished with regard to the environment. Begin by noting some of the environmental dangers which we face; a lack of resources and a lack of space to put waste products (compounded by a rapidly growing population); energy sources that generally use fossil fuels and pollute the environment; manufacturing processes that are complex, costly and often polluting; and an apparent increase in the worldwide greenhouse effect, due in part to a loss of forest land. About half of our solid waste is paper, and recycling five to ten tons of paper saves about a million trees! Explain to the students that most areas of North America now require recycling of some kind. Usually paper and metals are the easiest materials to recycle.

BE SAFE! For this activity, **supervise the use of any electrical appliances with care.** Students should work in groups of two or more.

GUIDE TO THE INVESTIGATION:

Students will need:

> old newspaper
> electric blender, or mixing bowl with electric mixer or egg beater
> cake pan, or other flat aluminum or glass pan
> wire screening (e.g., window screening) to cover cake pan
> water
> cornstarch, or wallpaper paste or flour
> stirrer (spoon or stirring rod)
> wax paper
> rolling pin, or large glass jar

Have the students carry out the activity. Review the results.

You might choose to use several of the further study suggestions as a basis for a "make the best paper" contest.

Note that the latest thoughts on possible government assistance with building a recycled products market is through tax incentives and bidding priority for companies using such products.

VOCABULARY: ecology, recycling, pollution

ADDITIONAL RESOURCES: A visit to a local recycling plant is a fascinating adventure. Prepare the students in advance with questions, including some which they generate. A subsequent visit to a paper mill provides for a dramatic opportunity to compare the processes and costs of using new materials with those of recycling.

In addition, the Innovation broadcast:"Down in the Dumps," is excellent. Contact a PBS station to find out how to obtain a copy.

TEACHER'S GUIDE
SOLAR COLLECTORS

GOAL: To develop student understanding of methods of collecting solar energy.

STUDENT OBJECTIVES:
- To comprehend the importance of solar energy.
- To build and test variations on solar collectors.

PRELAB DISCUSSION: The earth appears to be warming up, probably as a direct result of burning fossil fuels. Other factors, such as deforestation are at work as well. Alternative fuels must be found, both to reduce the greenhouse effect and to reduce pollution. Ironically, one form of an outstanding alternative supply of energy, solar energy, relies on the greenhouse effect in order to work. Students should investigate the present energy situation before starting this activity. They should be familiar with the greenhouse effect. A class discussion on how to capture the sun's energy best also would enhance the investigation. The apparatus suggested here is designed to be modified. Students have painted the insides different colors; have used wavy, rather than flat inside surfaces in order to increase surface area; have used external mirrors to help add to the incident light; have tried different insulation materials; and in longer-term projects, have built two-level top surfaces of glass and tried different combinations of greenhouse gases between them. Students are limited only by the time available to devote to this activity, the materials available and safety factors. Contests to build the best collector are fun and worthy of positive public relations. This is an area in which average students can come up with original and workable ideas. Students should work in groups of two or more.

BE SAFE! **Goggles should be worn. Hoses should be positioned so that no blockages occur.**

GUIDE TO THE INVESTIGATION:

Students will need:

    wood (or a large shoe box)
    tools
    glue or nails

insulation material
heat lamp (or the sun)
plastic wrap
books to elevate the collector
2 thermometers
2 buckets
2-3 meters of clear, thin tubing
materials to vary the experiment as desired
black paint
paint brush
tape
meter stick
watch or other timer
liter beaker
large beaker with bottom spout
screw clamp for the hose
goggles

Supervise the construction and testing of the collectors. Approve any original designs before construction begins. Discuss the results. Concentrate on ways to make the collectors even more effective.

VOCABULARY: fossil fuel, greenhouse effect

ADDITIONAL RESOURCES: The New York State Energy Research and Development Authority has run a Student Energy Research Competition each year for the past eight years. Call them (toll free) in Albany, New York, for lists of the 100 projects awarded grants each year. Most are creative rather than complex.

It is helpful and positive to bring in "experts" from the school system to help with this activity. For example, social studies people can report on the social and political consequences of pollution, while industrial arts people can help teach the students about skills of design and construction.

TEACHER'S GUIDE
VOLUME, AREA AND SOLAR HOMES

GOAL: To understand important factors in an energy efficient home.

STUDENT OBJECTIVES:
- To investigate the energy losses in various volume to area configurations.
- To explain several factors necessary to an energy efficient home.
- To build and test models of passive and/or active energy efficient solar homes.

PRELAB DISCUSSION: This is the second of four activities dedicated to the recognition of the need for greater energy efficiency and to the beginning of a search for solutions to the problem. Begin with a discussion of the world's energy shortage. Point out the high cost of energy and the fact that if all of the world's people lived as do Americans, there would not be enough available energy today in the entire world. This is a cause of international tension when Americans ask other countries to do things, such as stop cutting rain forests, to preserve the environment. They tell us to stop wasting energy! If all are to live in an environmentally sound world, all will have to live more efficiently. Ask the students how energy is wasted in the home. You may choose to have them conduct a formal energy survey, including physical aspects of their homes (insulation, window types and numbers) and family habits (use of heat and electricity). The parts of this activity are independent of each other. The first part investigates the fact that heat is lost most rapidly out of long, flat structures, and least out of spherical ones (followed by large cubes). Help the students calculate the volume to surface area ratio. Any shapes which can contain a liquid may be used. The key is that they should be made of the same material and have the same wall thickness. If Pyrex containers are not available, different height-to-width ratio metal cans are obtained most easily. If other materials are used, introduce hot, rather than boiling, water.

BE SAFE! **The teacher should boil and pour the water! When using hot water students should wear goggles and a lab apron and should exercise safety precautions.**

In the next part of the activity, the limiting factor is simply the time which may be devoted to the project. The project can range from a simple, shoe box model energy efficient home to a true scale model incorporating many aspects of the present technology. You probably will wish to advise the students as to the time available for their investigation. Start the students thinking with a discussion of which way a house should face with respect to the wind and how (deciduous) trees can be used to shade areas in summer and allow sunshine in winter. Direct the students to appropriate resources. Students should work in groups of two or more.

GUIDE TO THE INVESTIGATION:

Students will need:

Part 1:

At least two of the following (2 cubes can work):

> 1 to 8 Pyrex cubes (see discussion)
> Pyrex cylinder (same volume as cube, if possible)
> Pyrex sphere (same volume as cube, if possible)
> Pyrex rectangle (same volume as cube, if possible)

BE SAFE!

> boiling water (**poured by the teacher**)
> thermometer(s)
> timer or clock
> metric ruler
> goggles
> laboratory apron

Part 2:

> Shoe box or wood, tools and nails
> paper, cardboard, and tape
> ring stand and pole
> 150 watt lamp and base
> thermometer
> timer or clock
> metric ruler
> other materials, as needed

VOCABULARY: active solar home, conservation, insulation, passive solar home, solar collector

ADDITIONAL RESOURCES: Utilities offer (usually free) energy surveys of homes, including recommendations for improvements. This should be done for all homes. There are hundreds of books about the energy efficient solar home. Their ideas range from the simple up to very sophisticated heat pumps. The best way to explore this topic is to visit a highly efficient solar home. Check your area for any which have been constructed for research at a laboratory or university. Brookhaven National Laboratory on Long Island, for example, has constructed and tested an extraordinarily efficient solar home.

TEACHER'S GUIDE
THE BEST LIGHT BULBS

GOAL: To increase student knowledge of conservation by investigating the efficiency of light bulbs.

STUDENT OBJECTIVES:
- To become familiar with the basic terms used in light bulb comparisons.
- To measure the relative efficiencies of different light bulbs.

PRELAB DISCUSSION: Discuss the need for energy conservation with the students. Reinforce the idea that conservation can be achieved through increased efficiency, and does not have to mean that sacrifice is necessary. Today, light bulb packages are marked with their power expressed in luminous flux, measured in lumens. Just tell students that lumens compare the amount of light falling at a distance one meter from the bulb. (Actually, 12.56 lumens is equal to the light emitted from a one candela source of luminous intensity. The lumen takes into account the fact that the eye is not equally sensitive to all frequencies of the power output of a light bulb. Thus, green light may be far more useful to the eye than the red or violet. A lumen of USEFUL power may come from less than .002 watts of green light, but much more from the less efficiently used red and violet.) Students should be familiar with how light decreases with distance. If treated mathematically, the illumination on an object, **E**, falls off as the inverse square of the distance from the emitting source.

$$E = \text{lumens}/d^2$$

In non-mathematical terms, if you go twice as far from a bulb you get a quarter the light; three times as far you get 1/9 the light, and so on.

BE SAFE! **Be sure to review electric safety procedures. Include the fact that no one should ever touch any exposed wires.**

GUIDE TO THE INVESTIGATION:

Students will need:

    two large paraffin blocks about 10 x 15 x .5 cm
    aluminum foil
    scissors

clear tape,
meter stick
2 identical new 100 watt light bulbs with two sockets
bulbs of different wattage (fluorescents are good to test!)
G.E. Miser bulb with the same lumen rating as another of the bulbs
bulb of equal wattage but a different lumen rating than another

This experiment has been designed for simplicity of materials. Any photometer can be used to replace the paraffin blocks suggested here, and the light bulbs can be mounted on standard meter stick supports, but it is not necessary to do so.

Students should carry out the activity. Results and sources of error should be analyzed.

VOCABULARY: illumination, lumen

ADDITIONAL RESOURCES: General Electric and Westinghouse research facilities are cooperative in discussing how they investigate new ideas in lighting. In addition to efficiency, cost of production and aesthetic appeal are other factors influencing research.

TEACHER'S GUIDE
INVESTIGATING INSULATION

GOAL: To develop student understanding of energy conservation.

STUDENT OBJECTIVES:
- To recognize that heat flows from areas of high temperature to areas of low temperature.
- To discover that different materials have widely differing insulating properties.

PRELAB DISCUSSION: This is a good experiment in which to apply the personal experiences of students to a scientific discussion. Students should be asked about the energy crises of the 1970's and why there appears to be no crisis now. Energy conservation should be defined. Students can be asked how their home is insulated. After the experiment is completed, students can discuss additional ways in which energy can be conserved in their lives. If necessary, appropriate graphing techniques should be reviewed. For this activity, students should work in groups of two or more.

BE SAFE! Students should wear goggles during this experiment. Caution is necessary when handling hot water.

GUIDE TO THE INVESTIGATION:

Students will need:

Supply of boiling water. Each group will need a measured quarter of a liter (use a 400 ml or larger Pyrex beaker) three times, at approximately six minute intervals. **The water should be poured by the teacher.**

BE SAFE! Care must be taken when distributing and pouring boiling water. Goggles and lab aprons must be worn.

- 3 identical pop-top soda cans
- boiling water
- clock, watch or timer
- Celsius scale thermometer
- metric ruler
- cardboard
- scissors
- masking tape
- goggles
- lab apron

BE SAFE! insulating materials such as cardboard, newspaper, carpet remnants, Styrofoam, fiberglass insulation (**handle with proper gloves**), wool or other cloth, or other items (**no asbestos!**)

After the prelab discussion, ask students to select two materials to test. Each group will cover the sides of two cans with their "insulation." In order to make accurate comparisons, the cans should be covered to the same thickness: approximately one centimeter. The third can is the control and remains uninsulated. Cardboard tops for the cans should be cut. A thermometer hole should be included.

Place 250 ml of boiling water in one can for each group. Have students cover the can, stabilize the thermometer, and record data for six minutes. Repeat with the other two cans.

Have students graph the data and write a conclusion. Assist students with the graphing.

Discuss the results. A nice comparison can be made by having each group graph their data on an overhead transparency with pre-drawn equal axes. The class results can be compared simultaneously by layering the transparencies on an overhead projector.

VOCABULARY: conservation, insulation, R-value

ADDITIONAL RESOURCES: Local utilities offer much free information on conserving energy. A very nice "Insulation Materials Board" displaying and describing numerous insulation products is available from Frey Scientific Company (905 Hickory Lane, Mansfield, Ohio 44905).

TEACHER'S GUIDE
AIR POLLUTION AND YOU

GOAL: To comprehend the nature and extent of the air pollution problem.

STUDENT OBJECTIVES:
- To describe some sources of air pollution.
- To recognize carbon dioxide as a major greenhouse gas.
- To collect and observe particulates.
- To discover some effects of acid rain.

PRELAB DISCUSSION: This activity is designed to survey the major air pollution problems. Begin with a discussion about the negative effects of dirty air on health. Discuss the major sources of air pollution, and note that some of them are natural. It should be noted that every aspect of air pollution can be investigated in great depth, and that the laboratory work discussed here can be extended easily, if desired. Students should work in groups of two or more for this activity.

BE SAFE! **Goggles and a lab apron should be worn. Part 3 should be done as a demonstration!**

GUIDE TO THE INVESTIGATION:

Students will need:

Part 1: carbon dioxide

beaker	goggles
straw	laboratory apron
limewater	

Part 2: particulates

microscope	goggles
slides	laboratory apron
petroleum jelly	

Part 3: acid rain (demonstration only!)

6 test tubes	nylon
test tube rack	aluminum

sulfuric acid
marble chips
wool
cotton

lead
goggles
laboratory apron

Conduct the parts of this activity. Review the results.

Be certain to discuss the positive things that are occurring in the field of air pollution control. For example, there are laws requiring cars and industry to reduce the amount of pollutants which they emit. However, on the negative side, as the population has increased the global demand for more cars, living space and products seems to have countered the effects of the present laws.* "What steps to take to reduce air pollution?", makes an outstanding topic for a debate, as the steps may prove costly and may interfere with individual freedoms such as the right to drive when and where a person likes. Be certain to point out to students that smoking is a major personal source of air pollution.

* Two studies, conducted in the late 1980's by the American Museum of Natural History and by the United Nations Environmental Programme, found a nearly perfect correlation between human population growth and the greenhouse gases carbon dioxide (past 25 years) and methane (past 500 years).

VOCABULARY: acid rain, greenhouse effect, particulates, pollution

ADDITIONAL RESOURCES: An environmental chamber permits sophisticated examination of the effects of different air and water pollutants on materials and on living things. A good chamber is available from Hubbard/Jewel (P.O. Box 104, Northbrook, Illinois 60062). Be certain to use current events materials when studying pollution. In addition, there are wonderful speakers, films and television shows on the topic of air pollution. The television show, Innovation, has had several, including "Air Pollution: Outdoors," and "Air Pollution: Indoors." Contact a local PBS station to find out how to obtain a copy.

TEACHER'S GUIDE
WATER POLLUTION AND WATER PLANTS

GOAL:

To comprehend the nature and extent of the water pollution problem.

STUDENT OBJECTIVES:
- To describe some sources of water pollution.
- To discover some effects of water pollution on water plants.

PRELAB DISCUSSION:

This activity is designed to investigate a few of the major water pollution problems. Begin with a discussion about the negative effects of contaminated water on health. Discuss the major sources of water pollution. It should be noted that every aspect of water pollution can be investigated in great depth, and that the laboratory work discussed here can be extended easily, if desired. Students should work in groups of two or more.

BE SAFE! Goggles and a lab apron should be worn.

GUIDE TO THE INVESTIGATION:

Students will need:

> 7 jars with lids (about a liter in size)
> marking pencil
> aquarium or pond water
> duckweed, or elodea or other water plants
> teaspoon
> lawn fertilizer
> powder, phosphate based detergent
> powder, non-phosphate detergent
> liquid detergent
> liquid soap
> motor oil
> goggles
> lab apron

Have the students carry out the procedures. Review the results.

Be certain to discuss the positive things that are occurring in the field of water pollution control. For example, there are laws requiring the treatment of liquid waste and sewage, and the removal of toxic materials, before placement into water systems.

VOCABULARY: pollution, toxic

ADDITIONAL RESOURCES: An environmental chamber permits sophisticated examination of the effects of different air and water pollutants on materials and on living things. A good chamber is available from Hubbard/Jewel (P.O. Box 104, Northbrook, Illinois 60062). Be certain to use current events materials when studying pollution. In addition, there are wonderful speakers, films and television shows on the topic of water pollution. Your local water department or state wildlife department usually provides speakers on this topic.

TEACHER'S GUIDE
A TRUE SCALE MODEL OF OUR SOLAR SYSTEM

GOAL: To have students understand the truly immense size of our solar system.

STUDENT OBJECTIVES:
- To observe a standard model of the solar system.
- To be able to calculate map road sizes using a scale.
- To calculate the sizes of the planets and their distances to the sun using a scale.
- To build a true scale model of the solar system using this scale.

PRELAB DISCUSSION: If you have a model of the planets, or of the sun, moon and earth, bring it out as an introduction to this activity. If not, use the illustration in the student section. Some students will need help with the calculations. Doing one or two samples and then giving the students the rest of the answers is appropriate for many students. You will need a very long outside area to do this activity, but it is worth it. This is an activity which they remember for years! Take pictures!

GUIDE TO THE INVESTIGATION:

The class will need:

    aluminum foil   meter stick

If an outside area is not available, a long corridor can be used. It will accommodate at least the sun and first few planets.

Conduct the activity. Discuss the results. Stress the fact that the planets actually orbit around the sun and that in this experiment we looked at only one side of the sun. Encourage students to try a few of the suggestions at the end of the activity.

VOCABULARY: asteroid, comet, moon, orbit, revolve, satellite, scale, solar system

ADDITIONAL RESOURCES: An up-to-date reference for the solar system is Abell, George O., *Realm of the Universe*, New York, Saunders,

1988. There are hundreds of fascinating books on space travel, but for stimulation the NASA films on the flights of the Voyagers are wonderful. These may be obtained through NASA CORE, Lorain County JVS, 15181 Route 58 South, Oberlin, OH 44074, or by calling the National Aeronautics and Space Administration in Washington, DC.

TEACHER'S GUIDE
THE EARTH'S PATH THROUGH SPACE

GOAL: To develop student understanding of the complexities of the motions of heavenly bodies.

STUDENT OBJECTIVES:
- To trace the path of a point on earth as it goes through space.
- To comprehend the problems of calculating paths for space vehicles.

PRELAB DISCUSSION: This activity helps the students to understand why it is not a simple problem to send a rocket ship to Mars or Pluto. If students have done the activity: *A True Scale Model Of Our Solar System,* then begin by noting the vast regions of space where there are no planets. Discuss the fact that the relatively tiny planets are all rotating and revolving at different speeds, and that this makes it harder to calculate the path a rocket ship must take. Note, in addition, that the rocket ship must miss any moons and/or rings of a target planet, and must face different amounts of gravity than on earth. Finally, remind the students that only the size of the earth and moon are approximately to scale in the first part of this activity. This activity is done best in a large indoor space, such as a gymnasium. A classroom may be used if the desks and chairs are pushed to the outside walls.

GUIDE TO THE INVESTIGATION:

Each class will need:

> white paper
> 30 cm diam. globe or beach ball
> baseball or tennis ball
> tongue depressor (or piece of wood)
> aluminum foil
> scotch tape
> marking pen
> meter stick

Conduct the investigation and discuss the results.

VOCABULARY: moon, orbit, revolve, rotate, scale, solar system

ADDITIONAL RESOURCES: An up-to-date reference for the solar system is Abell, George O., *Realm of the Universe*, New York, Saunders,

1988. There are hundreds of fascinating books on space travel, but for stimulation the NASA films on the flights of the Voyagers are wonderful. These may be obtained through NASA CORE, Lorain County JVS, 15181 Route 58 South, Oberlin, OH 44074, or by calling the National Aeronautics and Space Administration in Washington, DC.

TEACHER'S GUIDE
A BALLOON MODEL OF THE UNIVERSE

GOAL: To comprehend the concept of the expanding universe.

STUDENT OBJECTIVES:
- To describe the evidence for an expanding universe.
- To build a balloon model of the expanding universe.
- To recognize the limitations of the model.

PRELAB DISCUSSION: Students should be familiarized with concepts such as galaxy and light year. It is helpful to have conducted the activity: *A True Scale Model Of Our Solar System*, as this will give the students a feeling for the vastness of space. This is an easy activity to conduct, and leads to interesting discussions about the creation of the universe. Students should work in pairs.

GUIDE TO THE INVESTIGATION:

Students will need:

- balloon
- pen
- metric ruler
- string
- scissors

Students should complete the procedure section of the activity. Allow adequate time to discuss the results in terms of possible origins, shapes and destinies of the universe.

VOCABULARY: big bang theory, galaxy, light year, universe

ADDITIONAL RESOURCES: The work of Edwin Powell Hubble, the theories of the origin of the universe and the evidence for each are discussed extensively in: Abell, George O., *Realm of the Universe*, New York, Saunders, 1988. In addition, an extraordinary view into current research in this field is presented in: Hawking, S. W., *A Brief History of Time*, New York, Bantam Books, 1988.

TEACHER'S GUIDE
SCIENCE JEOPARDY

GOAL: To develop an instrument which will provide an exciting format for study and review of scientific concepts and terminology.

STUDENT OBJECTIVES:
- To develop good study skills by learning to group ideas into topics for study.
- To compete in a contest involving their studies.

PRELAB DISCUSSION: Discuss the idea that good study habits take practice to develop, just as good sports skills do. Tell the students that they are going to compete in a Science Jeopardy game, modeled on the television show Jeopardy. You may want to tell students which topics to review in advance of the game, or you can choose to surprise them. You can use the game to motivate the students to review, or to have individual or team testing take place. The game is easily adapted for students of all ability levels. Of course, this can be a great way to simply have fun or to use as a class reward. Enjoy!

GUIDE TO THE INVESTIGATION:

Students will need:

  Science Jeopardy answers

Review the rules of the game. They are stated in the *PROCEDURE* section. Any science topic at all may be used as a source for questions. Hint: type the questions on index cards, put the point values on the other sides, and place them face down on the desk in front of you. Put the Science Jeopardy game board, as shown here, on the chalkboard. Fill in the category names and the point values. As students select a square, simply cross out its point value on the board. Record each individual's or group's score on the board. Set a time limit for Single Science Jeopardy, a similar time limit for Double Science Jeopardy, and a limit for a Final Science Jeopardy question, if one is used. Set a time limit for each question, perhaps thirty seconds. Do not allow students to interrupt the time of other students.

VOCABULARY: (As appropriate)

ADDITIONAL RESOURCES: Sources of questions may include your own notes, the student textbook, other books, teacher resource guides and games, such as Trivial Pursuit, which have science as a category.

GAME BOARD LAYOUT: SINGLE SCIENCE JEOPARDY				
CATEGORY 1	CATEGORY 2	CATEGORY 3	CATEGORY 4	CATEGORY 5
100	100	100	100	100
200	200	200	200	200
300	300	300	300	300
400	400	400	400	400
500	500	500	500	500

For the Double Science Jeopardy board, simply double each point value listed above.

SAMPLE ANSWERS AND QUESTIONS: CATEGORY: PLANTS

100 These Draw Up Water From The Ground
    (WHAT ARE ROOTS?)

200 Two Things In Plant Cells But Not Animal Cells
    (WHAT ARE CELL WALLS AND CHLOROPLASTS?)

300 Green Chemical Used By Plants To Help Make Food
    (WHAT IS CHLOROPHYLL?)

400 Stomates
    (WHAT ARE OPENINGS IN LEAVES?)

500 Movement Of Materials Across A Membrane From Areas Of High To Low Concentration
    (WHAT IS OSMOSIS?)

TEACHER'S GUIDE
SCIENCE BINGO

GOAL: To develop an instrument which will provide an interesting method for study and review of scientific concepts and terminology.

STUDENT OBJECTIVES:
- To develop good study skills by learning to group ideas into topics for study.
- To compete in a contest involving their studies.

PRELAB DISCUSSION: Discuss the idea that good study habits take practice to develop, just as good sports skills do. Tell the students that they are going to play Science Bingo. You can use the game to get students to review or to have individual or team testing take place. You may want to tell the class which topics to review in advance of the game. Of course, this can be a great way to simply have fun or to use as a class reward. Enjoy!

GUIDE TO THE INVESTIGATION:

Students will need:

    Science Bingo card
    tokens to cover the squares

    Suggested categories are:

        B for biology
        I for inventions
        N for names of scientists
        G for geography
        O for organs of the body

VOCABULARY: (As appropriate)

ADDITIONAL RESOURCES: Sources of questions may include your own notes, the student textbook, other books, teacher resource guides and games, such as Trivial Pursuit, which have science as a category. It can be fun to have students develop a pool of questions for homework or for extra credit. Another nice variation is to use a student as the moderator.

TEACHER'S GUIDE
SCIENCE TAG TEAM RACE

GOAL: To develop an instrument which will provide an exciting format for study and review of scientific concepts and terminology.

STUDENT OBJECTIVES:
- To develop good study skills by learning to group ideas into topics for study.
- To compete in a contest involving their studies.

PRELAB DISCUSSION: Discuss the idea that good study habits take practice to develop, just as good sports skills do. Tell the students that they are going to compete in a Science Tag Team race. You may want to tell the class which topics to review in advance of the game. You can use the game to motivate the students to review, or to have individual or team testing take place. Of course, this can be a great way to simply have fun or to use as a class reward. Enjoy!

GUIDE TO THE INVESTIGATION:

Students will need:

identical sets of ten cards, one set per group.

The cards may cover any topic at all. Number each card. Each identical question should receive the same number. Hint: type the questions on paper, appropriately spaced and sized, reproduce the paper, and cut and paste each question onto index cards. A sample set for the topic of weather is included in this activity.

Organize the students into standing lines or seated rows of ten or fewer students. For example, if the class has twenty-four desks, you may choose to line up the desks into four rows of six, and have the students remain in seats. Assign a number to each row.

Shuffle each set of cards and place a set, face down, at the front of each row. Divide the chalkboard into as many sections as there are teams. Put a team number at the top of each section.

When you say "Go," the first person on each team picks up the top card from the pile, goes to the board, puts down the question number and their answer. They then go back, tag the desk or hand of the next player on their team, and sit down or get at the end of the line. When a team has completed ten questions and is back in place, the game is over. Each team is given ten points for a correct answer. If you wish, you may play several rounds in one day. You and the children will certainly have fun with this!

VOCABULARY: (As appropriate)

ADDITIONAL RESOURCES: Sources of questions may include your own notes, the student textbook, other books, teacher resource guides and games, such as Trivial Pursuit, which have science as a category.

TEACHER'S GUIDE
AN EGG DROP CONTEST

GOAL: To develop a packaging material, using scientific reasoning, so that eggs will not break when dropped in the material from a height.

STUDENT OBJECTIVES:
- To investigate qualities of packing materials.
- To compete in a contest in which eggs are dropped in order to test chosen packing materials.

PRELAB DISCUSSION: Discuss the idea that scientific reasoning can and should be used to help find answers to everyday problems. For example, knowledge of the way things travel through air, accelerate when dropped and bounce when they hit can aid in the design of packaging. Talk about how the actions are similar, but different, when a hardball hits a bat, a tennis ball hits a wall or a sponge is dropped to the floor. Organize a time and location for a contest, allowing students time to conduct research and testing. The testing can be done at home or after school in a supervised location. Each student or group should keep their design a secret until the contest. Take pictures and have fun!!

BE SAFE! Remind students that they should not climb to heights to test their packaging.

GUIDE TO THE INVESTIGATION:

Students will need:

>   shoe box   stopwatch
>   building brick (20 x 10 x 5 cm)   meter stick
>   two raw eggs   masking tape
>   various types of packing material (gelatin, pop-corn, newspaper, etc.)

Give the students the goals and rules for the egg drop contest. They must follow the instructions for the arrangement of the box and eggs, as shown in the *Procedure* section. Give them a date, time and location for the contest and help direct them to the kind of packaging to look for. Select a date for submission of their secret packaging specifications for your approval.

BE SAFE! The packaged eggs should be dropped **BY THE TEACHER** from a height of 10 meters, approximately the height of a two or three story school building, onto a flat, concrete surface. **Use a lower height if there is a problem or safety difficulty. Students should stand well away from the drop zone.** They should be prepared to clean the results of any unsuccessful drops.

Review the qualities of the successful packages.

VOCABULARY: energy transfer

ADDITIONAL RESOURCES: The National Science Olympiad, as well as several college physics departments, now conduct similar contests. Visit a nearby contest. What were the characteristics of the winner?

Basic Physics Science Activities

by

Dr. C. V. Krishnan

TEACHER'S SECTION
MEASUREMENT OF LENGTH

GOALS AND OBJECTIVES:

1. To learn about the metric system of measurement.
2. To learn the use of a metric ruler to measure lengths.
3. To learn to measure lengths that have fractions of a centimeter.
4. To learn to convert millimeters into centimeters.
5. To learn to convert centimeters into meters.
6. To learn that the unit of length according to the International System of Units (SI) is a meter.

DISCUSSION BEFORE THE LABORATORY ACTIVITY:

It is very important to stress the need for wearing an apron and safety goggles.

Originally, the standard meter was defined as one ten–millionth of the distance of the pole of the earth to the equator along the meridian passing through Paris. This length was marked on a bar of platinum–iridium alloy kept at zero degree Celsius. Since the bar expands on heating, the temperature must be specified. This bar was kept at the International Bureau of Weights and Measures at Sevres, France. For practical use, copies of this bar were used. Later, the standard meter was defined in terms of the length of a wave of light, the orange light produced by krypton gas. More recently, the standard meter is defined as the distance traveled by light in 1/299792458 second. This is close to the distance traveled by light in one 300 millionth of a second.

One meter = 100 centimeters (cm). One centimeter = 10 millimeters (mm).

The students may be familiar with inches, feet, and yards. The advantages of using a metric system rather than the British system should be stressed. Most countries use the metric system for measurements. Show them a meter stick, a centimeter length and a millimeter length. Before starting this activity, students should practice to change millimeters into centimeters and centimeters into meters. They should recognize that one meter is 100 times bigger than a centimeter and that a centimeter is 10 times bigger than a millimeter.

Practice measuring the lengths of a variety of items such as a table, a book, a finger, a tile, the height of a person, the length of a room, and a candy. Challenge them to find the thickness of a sheet of paper, and the thickness of a piece of hair.

We measure very long distances such as the distances to other planets and very small distances such as the diameter of an atom and a nucleus.

TEACHER'S SECTION
MEASUREMENT OF VOLUME

GOALS AND OBJECTIVES:

1. To determine the volumes of solids that have regular shapes.//
2. To learn that the volume of an object is the same as the volume of water displaced by the object.
3. To learn that the volume of a large object is the same as the sum of the volumes of several pieces obtained by dividing the large object into several pieces.
4. To learn that the volumes of objects that have regular shapes can be computed using mathematical formulas.
5. To learn that the volume of an object determined by water displacement technique is the same as the volume determined by using an appropriate mathematical formula.
6. To learn that the volume of a liquid can be determined by measuring the amount of space it occupies in a graduated cylinder.
7. To learn that the volume of a gas in a container is the same as the volume of water occupying the same space in the container.
8. To learn that the volume occupied by a gas is the same as the volume of water it can displace from a container.

DISCUSSION BEFORE THE LABORATORY ACTIVITY:

It is very important to stress the need for wearing an apron and safety goggles.

Discuss the meaning of volume of an object. Discuss the SI unit of volume, cubic meter.

Show several objects to students and ask them to predict the comparative volumes of the objects. Ask them to suggest ways of checking out their predictions.

Introduce the concept of area of a square, and a rectangle. Then introduce the concept of volume of a cube and a rectangular solid.

Discuss the advantages of water displacement technique for measuring volumes. Challenge the students to design an apparatus for measuring the volume of a large object using water displacement technique.

Volumes of only solids that are not porous and that do not dissolve in water can be measured by water displacement.

Discuss the accuracies of measuring volumes using graduated cylinders of different sizes, graduated beakers, and baby nursing bottle (baby nurser). There are metric measurements in the baby nurser which can be conveniently used for measuring volumes.

Let the students practice cutting exact cubes and rectangular solids using substances such as apples and potatoes. This needs patience and extreme care. Cubes and rectangular solids of wood, plastic, and metals are also commercially available.

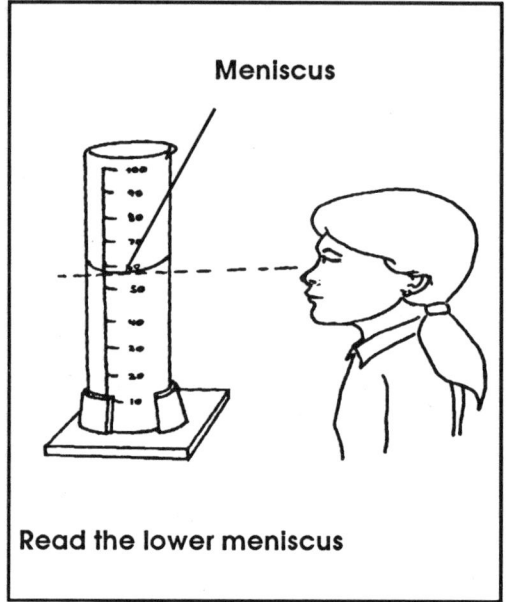

Correct way of reading the level of liquid in a graduated cylinder

Students may be familiar with, pints, quarts and gallons. Discuss the advantages of using milliliters and liters as the units for volume measurements. One liter = 1000 milliliters (mL). One milliliter and one cubic centimeter (cc) are nearly the same. One cubic meter = 1,000,000 cubic centimeters.

Make sure that students learn to read the lower meniscus of the liquid inside the graduated cylinder as the correct reading for volume. (See illustration 2–1.) Graduated cylinders are normally marked in milliliters.

Stress the need for specifying the temperature in volume measurements. Show the volume change of a liquid by heating a known volume of water in a graduated cylinder to a higher temperature.

The volume occupied by a gas depends on the pressure of the gas and the temperature of the gas. Show the compression of a gas by applying pressure. Also show the expansion of a gas by heating. If there is a gas in a bottle at atmospheric pressure, its volume can be measured by filling the bottle with water and measuring the volume of water. There is no easy way of measuring the volume of gas in a balloon. However, the gas from the balloon can displace water from a graduated cylinder and this displacement can be measured. This neglects the volume changes associated with the differences in the pressure of the gas inside the balloon and the graduated cylinder. The pressure of the gas inside a balloon depends on the extent to which the balloon is inflated. The volume of a gas depends on the pressure and the temperature. When the gas is collected inside a gratuated cylinder by water displacement, the pressure inside is equal to the atmospheric pressure when the levels of water inside the graduate and outside the graduate are the same.

TEACHER'S SECTION
MEASUREMENT OF MASS

GOALS AND OBJECTIVES:

1. To learn the use of a double-pan or beam balance for measuring masses.//
2. To learn to level the beam of a double-pan balance.
3. To learn that two similar-looking objects may not weigh the same.
4. To learn that the mass of an object depends on how much matter it contains in a given space.
5. To learn that the mass of an object depends on the nature of matter it contains.
6. To learn that real measurements give better information than from guessing.
7. To learn that the basic unit of mass according to the International System (SI) of Units is a kilogram.
8. To learn to change grams into kilograms.

DISCUSSION BEFORE THE LABORATORY ACTIVITY:

It is very important to stress the need for wearing an apron and safety goggles.

It should be emphasized that no eating is allowed during the hands-on activity. The materials used in this laboratory activity are common household food items. However, children should learn to treat these materials as laboratory items and no laboratory item should be eaten or tasted.

All balances are costly items, and children should learn to handle them carefully. Also, the balances are sensitive to small differences in masses, and it is important to hold the balance only at the base if it has to be removed from one place to another place.

Discuss the reasons the masses of an apple and an orange of similar size may be different.

The mass of an object depends on the quantity of matter it contains. The mass of an object is determined by comparing its mass with an object of known mass. A kilogram (kg) is the standard unit of mass according to the International System of Units (SI). It is based on the mass of a cylinder made from an alloy of platinum and iridium, which is kept in the Bureau of Weights and Measures at Sevres, France. A kilogram is divided into 1000 smaller units. This smaller unit is called a gram (g). A gram is further divided into 1000 still smaller units, called milligrams (mg).

1 kg = 1000 grams. 1 g = 1000 milligrams.

Only copies of the original 1 kilogram mass cylinder are available for use. Most countries measure masses in kilograms. This unit is much more convenient than pounds. One pound is approximately 453 grams.

Challenge students to visit a supermarket and write the masses of 10 familiar items in grams as well as in ounces or pounds. Then practice with them to change the masses into kilograms.

Let the students practice measuring the masses of small and large objects. Examples include a safety pin, a paper clip, a sheet of paper, a book, a piece of rock, and a tennis ball.

We measure huge masses such as the mass of the earth and small masses such as the mass of an electron.

TEACHER'S SECTION
MEASUREMENT OF TIME

GOALS AND OBJECTIVES:

1. To learn the use of stop–watches with split–intervals.

2. To learn to measure fractions of a second.

3. To learn that a ball drops to the floor from a height of one meter in less than a second.

4. To learn that the basic unit of time according to the International System of Units (SI) is a second.

DISCUSSION BEFORE THE LABORATORY ACTIVITY:

It is very important to stress the need for wearing an apron and safety goggles.

Stopwatch

The SI unit of time is a second. (See illustration 4–1.) Originally, it was defined on the basis of a mean solar day (the average time in a year taken by earth to make one complete turn on its axis). The day was divided into 24 hours. One hour = 60 minutes, and one minute = 60 seconds. The modern definition is based on an atomic clock that uses the vibrations of a particular kind of cesium atoms. This cesium atom has a mass of 133 atomic mass units. The time taken by a cesium–133 atom to make 9,192,631,770 vibrations is one second.

The SI units of length and mass use a decimal system. However, the SI unit of time does not use a decimal system. Practice with the students how to read the clock correctly. Compare the advantages of a digital watch over a clock with a dial.

Caution students to handle the Cronus split–interval stopwatch carefully because it is expensive. Practice with the students how to operate the stopwatch.

Approximate short intervals of time can be kept by learning to count numbers systematically. A sand clock also can keep short intervals of time. The sun and a sundial also can be used as clocks. The motion of the earth around the sun can be used for measuring long period of times. A stopwatch or a clock with a second hand can be used to measure times in seconds.

TEACHER'S SECTION
MEASUREMENT OF HEAT

GOALS AND OBJECTIVES:

1. To learn that heat lost by a given quantity of hot water is equal to the heat gained by another quantity of cold water.
2. To learn to use a mathematical formula for computing the number of joules of heat lost or gained.
3. To learn that heat flows from a body at higher temperature to a body at a lower temperature.
4. To learn that the SI unit of heat energy is joules.
5. To learn that heat is conserved.

DISCUSSION BEFORE THE LABORATORY ACTIVITY:

It is very important to stress the need for wearing an apron and safety goggles.

Discuss heat energy and the flow of heat energy.

Discuss the meaning of temperature.

Discuss the definitions of calorie, joule, and specific heat.

Discuss the formula for the number of joules.

Discuss exothermic and endothermic processes.

Temperature is a measure of the average kinetic energy of the molecules. Kinetic energy is the energy due to the motion of molecules.

Heat is a form of energy. A body at a higher temperature can transfer heat energy to a body at a lower temperature. When a body transfers heat energy, it loses some of its internal energy.

One calorie is the amount of heat needed to raise the temperature of 1 gram of water by 1 degree Celsius. The exact definition of calorie specifies the temperature of the water.

This calorie is different from the food calorie. One food calorie = 1,000 calories = 1 kilocalorie or 1 kcal.

Nowadays scientists use another unit, joule, the SI unit of energy. This is a derived unit. One joule is equal to the amount of work done by a force of one newton acting over a distance of one meter. Newton is the SI unit of force. One newton is the force needed to accelerate a mass of one kilogram by one meter per second per second because acceleration is meter per second per second. Work and energy have the same units. One joule = 4.18 calories.

Specific heat is the heat required to raise the temperature of 1 gram of a substance by 1 degree Celsius. Specific heat of water = 4.18 joules per gram per degree Celsius. The SI unit of mass is kilogram. If this unit is used, then the unit used for specific heat is kilojoules per kilogram per degree Celsius.

In an exothermic reaction or process, energy is released as heat. This energy is coming from the stored energy in the substance.

In an endothermic reaction or process, energy is absorbed and stored as potential energy.

Energy is conserved in all reactions and processes except in nuclear reactions. When energy is lost by a substance, it is gained by some other substance in joules per gram per degree Celsius.

Energy lost by a substance in joules = amount of substance (in grams) x change in temperature of the substance (in °C) x specific heat of the substance (in joules per gram per degree celsius).

310 Fundamentals of Science Activity Series

TEACHER'S SECTION
MEASUREMENT OF FORCE

GOALS AND OBJECTIVES:

1. To learn the difference between mass and weight.
2. To learn that the SI unit of force is a newton.
3. To learn that the force needed to lift an object is the same as its weight.
4. To learn that weight is a force and that the weight of an object is obtained by multiplying its mass in kilograms by the acceleration due to gravity in meters per second per second.
5. To learn that the force needed to lift an object increases with increasing mass of the object.

DISCUSSION BEFORE THE LABORATORY ACTIVITY:

It is very important to stress the need for wearing an apron and safety goggles.

Discuss matter and the meaning of mass of an object.

Discuss how to find the masses of different objects using a balance.

Discuss how to convert the mass of an object in grams into kilograms. One kilogram = 1,000 grams. One gram = 1,000 milligrams. If the mass of an object is 259 grams, then its mass also equals 0.259 kilogram.

Discuss force. A force is a push or a pull that will change the motion of a body. Only an equal and opposite push or pull can stop this motion.

The SI unit of force is a newton. One newton is the force needed to accelerate a mass of one kilogram by one meter per second per second.

$F = ma$, where F is the force in newtons, mass is the mass of the object in kilograms, and a is the acceleration in meters per second per second. The unit for acceleration is written as meters per second per second or ms^{-2} or m/s^2.

Discuss weight. We often use the term weight erroneously for mass. Mass and weight are different. Weight is measured using a spring balance. Spring balances are often calibrated in the unit for mass (kilogram or gram) as well as in the unit for force (newton).

In order to compute weight, one has to multiply mass in kilograms by acceleration due to gravity (9.8 m/s^2).

Gram	Newton
0	0
100	0.98
200	1.96
400	3.92
600	5.88
800	7.84
1000	9.80

Weight is measured in newtons. Weight is a force.

TEACHER'S SECTION
MEASUREMENT OF WORK

GOALS AND OBJECTIVES:

1. To learn the relationship between force and work.
2. To learn that the SI unit of work is a joule.
3. To learn that the work needed to lift an object can be computed by multiplying the weight of the object by the distance (in meters) it has to be lifted.
4. To learn that weight is a force and that the weight of an object is obtained by multiplying its mass in kilograms by the acceleration due to gravity in meters per second per second.
5. To learn that the force needed to lift an object can be reduced by using an inclined plane.
6. To learn that the force needed to lift an object can be reduced by reducing the angle of the inclined plane and by increasing the distance through which the force acts.
7. To learn that the work needed to move an object to a certain height is the same no matter how you move the object.

DISCUSSION BEFORE THE LABORATORY ACTIVITY:

It is very important to stress the need for wearing an apron and safety goggles.

Discuss force. A force is a push or a pull that will change the motion of a body. Only an equal and opposite push or pull can stop this motion.

The SI unit of force is a newton. One newton is the force needed to accelerate a mass of one kilogram by one meter per second per second.

$F = ma$, where F is the force in newtons, m is the mass of the object in kilograms, and a is the acceleration in meters per second per second. The unit for acceleration is written as meters per second per second or ms^{-2} or m/s^2. Another name for a newton is kilogram meter per second per second.

Discuss weight. We often use the term weight erroneously for mass. Mass and weight are different. Weight is measured using a spring balance.

Spring balances are often calibrated in the unit for mass (kilogram or gram) as well as in the unit for force (newton).

In order to compute weight, one has to multiply mass in kilograms by acceleration due to gravity (9.8 m/s^2).

Discuss work. In order to move any object from one place to another, we have to use a force. A force does work when it acts on an object and moves the object in the direction of the force.

Work = force x distance. The SI unit of work is the joule. This is a derived unit. One joule is equal to the amount of work done by a force of one newton over a distance of one meter. Another name for a joule is the newton meter.

Discuss the advantages of using an inclined plane to move a heavy object to a certain height. The force needed to move the object can be reduced by using the inclined plane. The smaller the angle of the inclined plane, the smaller the amount of force needed to move the object to the required height. However, the smaller the angle of the inclined plane, the greater the length of the inclined plane needed to reach the same height. The smaller force has to act over this extended length for moving the object. The amount of force needed to move the object can be measured using a spring balance calibrated in the unit of force.

The amount of work needed to lift an object to a certain height in air will be less than the amount of work needed to move the object to the same height using an inclined plane. This is because of the differences in friction exerted by air and the material surface of the inclined plane. However, the use of the inclined plane makes the force needed to move the object much smaller than that needed for lifting.

TEACHER'S SECTION
BOUNCING BALL

GOALS AND OBJECTIVES:

1. To learn the relationship between height and potential energy.
2. To learn the difference between potential energy and kinetic energy.
3. To learn that potential energy can be changed into kinetic energy, and back into potential energy any number of times.
4. To learn that a ball will stop bouncing off the floor after some time, because it loses some energy as heat to the surroundings.
5. To learn that the bouncing of a ball from the floor depends on the height from which it is dropped.
6. To learn how to draw a graph.

DISCUSSION BEFORE THE LABORATORY ACTIVITY:

It is very important to stress the need for wearing an apron and safety goggles.

Discuss potential energy. It is the stored energy in an object because of its position and because of the chemical make-up of the substance. Gasoline has a lot of potential energy because of its chemical composition. A rock on the top of a mountain has a lot of potential energy because of its position. The greater the height at which a body is kept, the greater its potential energy.

Potential energy due to position = mass x acceleration due to gravity x height of the body. Mass is in kilograms, acceleration is in meters per second per second; and height is in meters. The unit for potential energy is the joule.

Potential energy can be changed into work. Potential energy can also be changed into kinetic energy.

Discuss kinetic energy. It is the energy due to the motion of a body. Kinetic energy = 1/2 mass x velocity^2. The greater the mass and the greater the velocity of a body, the greater its kinetic energy. The unit for kinetic energy is the joule, when the mass is in kilogram and the velocity is in meters per second.

Kinetic energy can be changed into work. Kinetic energy can also be changed into potential energy.

When a ball is held at a certain height, it possesses potential energy because of its position. When it is allowed to fall, this potential energy changes into kinetic energy. When it hits the floor, it is completely changed into kinetic energy. When the ball bounces back, the kinetic energy changes back into potential energy. The ball does not hit the floor and bounce off the floor indefinitely because some energy is lost as heat to the surroundings. Thus the ball may not bounce back to the same height and eventually it will come to a stop.

Discuss how to draw a graph. Teach the terms X–axis and Y–axis for the coordinates of the graph. Use only a pencil to draw graphs on graph paper. Use a ruler to draw lines. Teach them to indicate the data points as small circles. Do not join point to point. Draw a line representing the data points. Discuss the reasons why some of the data points do not touch the straight line. Discuss the advantages of using a graph to interpret data.

TEACHER'S SECTION
BOUNCING OF DIFFERENT BALLS

GOALS AND OBJECTIVES:

1. To learn the relationship between height and potential energy.

2. To learn the difference between potential energy and kinetic energy.

3. To learn that potential energy can be changed into kinetic energy and back into potential energy any number of times.

4. To learn that a ball will stop bouncing off the floor after some time because it losses some energy as heat to the surroundings.

5. To learn that the bouncing of a ball from the floor depends on the chemical make-up of the ball as well as the nature of the surface of the floor.

DISCUSSION BEFORE THE LABORATORY ACTIVITY:

It is very important to stress the need for wearing an apron and safety goggles.

Discuss potential energy. An object contains a certain amount of stored energy because of its position and because of its chemical make-up. Gasoline has a lot of potential energy because of its chemical composition. A rock on the top of a mountain has a lot of potential energy because of its position. The greater the height at which a body is kept, the greater its potential energy.

Potential energy due to position = mass x acceleration due to gravity x height of the body. Mass is measured in kilograms, acceleration in meters per second per second, and height in meters. The unit for potential energy is the joule.

Potential energy can be changed into work. Potential energy can also be changed into kinetic energy.

Discuss kinetic energy. It is the energy due to the motion of a body. Kinetic energy = 1/2 mass x velocity^2. The greater the mass and the greater the velocity of a body, the greater its kinetic energy. The unit for kinetic energy is the joule.

Kinetic energy can be changed into work. Kinetic energy can also be changed into potential energy.

When a ball is held at a certain height, it possesses potential energy because of its position. When it is allowed to fall, this potential energy changes into kinetic energy, the energy due to motion. When it hits the floor, it is completely changed into kinetic energy. When the ball bounces back, the kinetic energy changes back into potential energy, the energy due to position. The ball does not hit the floor and bounce off the floor indefinitely, because it loses some energy to the surroundings as heat. Thus the ball may not bounce back to the same height and eventually it will come to a stop.

The chemical make up of different balls are different. Therefore they will bounce back from the floor to different extents. Also the same ball will bounce back to different extents depending on the nature of the surface of the floor.

TEACHER'S SECTION
THE NATURE OF THE FLOOR AND THE BOUNCING BALL

GOALS AND OBJECTIVES:

1. To learn the relationship between height and potential energy.
2. To learn the difference between potential energy and kinetic energy.
3. To learn that potential energy can be changed into kinetic energy and back into potential energy any number of times.
4. To learn that a ball does not bounce off the floor indefinitely because of some loss of energy as heat to the surroundings.
5. To learn that for a given ball, the extent of bounce from the floor depends on the nature of the surface of the floor.

DISCUSSION BEFORE THE LABORATORY ACTIVITY:

It is very important to stress the need for wearing an apron and safety goggles.

Discuss potential energy. It is the stored energy an object has because of its position and because of its chemical make–up. Gasoline has a lot of potential energy because of its chemical composition. A rock on the top of a mountain has a lot of potential energy because of its position. The greater the height at which a body is kept, the greater its potential energy.

Potential energy due to position = mass x acceleration due to gravity x height of the body. Mass is measured in kilograms, acceleration is in meters per second per second, and height is in meters. The unit for potential energy is the joule.

Potential energy can be changed into work. Potential energy can also be changed into kinetic energy.

Discuss kinetic energy. It is the energy due to the motion of a body. Kinetic energy = 1/2 mass x velocity^2. The greater the mass and the greater the velocity of a body, the greater its kinetic energy. The unit for kinetic energy is the joule.

Kinetic energy can be changed into work. Kinetic energy can also be changed into potential energy.

When a ball is held at a certain height, it possesses potential energy because of its position. When it is allowed to fall, this potential energy changes into kinetic energy. When it hits the floor, it is completely changed into kinetic energy. When the ball bounces back, the kinetic energy changes back into potential energy. The ball does not hit the floor and bounce off the floor indefinitely, because some energy is lost to the surroundings as heat. Thus the ball may not bounce back to the same height, and eventually it will come to a stop.

The chemical make–up of different balls are different. Therefore they will bounce back from the floor to different extents. Also the same ball will bounce back to different extents, depending on the nature of the surface of the floor.

Different kinds of surfaces are used for playing tennis in different tournaments. Even great players prefer a particular type of surface. They also experience difficulties in playing on some other surfaces because of differences in the bouncing of the ball.

TEACHER'S SECTION
FORCE DUE TO FRICTION

GOALS AND OBJECTIVES:

1. To learn that friction is a force.

2. To learn that the force due to friction opposes the motion of a body.

3. To learn that the force due to friction is measured in newtons.

4. To learn that the force needed to move an object increases with the increasing mass of the object.

DISCUSSION BEFORE THE LABORATORY ACTIVITY:

It is very important to stress the need for wearing an apron and safety goggles.

Discuss Newton's first law of motion. An object at rest remains at rest unless a force acts on the object to move it. A body in motion remains in motion in the same direction and at constant speed, provided there is no net force acting upon it to change the motion. Matter resists any change in its state of motion or rest. This property of matter is called inertia.

Discuss force. Force is a push or pull that changes the motion of an object. To stop this change in the motion of the object, another opposite and equal force of push or pull should act on the object.

Discuss friction. Show examples. Roll a rubber ball on the floor, and watch its straight line–motion. The ball gradually slows down and finally stops moving. Why didn't the ball continue to move at the same speed? Why did it violate the law of inertia? Ask the students why the ball slowed down and why it stopped moving after a while.

There is a net force acting on the ball. This force is not balanced by any other opposing force. The net force acting on the ball is the force due to friction. This force is exerted by the floor. The extent of this force depends on the nature of the surface of the floor. The surface of one object offers resistance to the movement of the surface of another object over it. This resistance of movement is called friction. Air also exerts a force due to friction. However, it is much less than the force due to friction exerted by the floor.

Measure the distance moved by the ball. Ask the students to suggest ways of making the ball move a longer distance by applying only the same initial force for starting the motion.

Why do we have to continuously supply gas to the engine to drive a car? (To continuously overcome the friction caused by the road and the air). Friction causes problems for machines. It causes machines to overheat and parts to wear out.

To start a surface sliding over another surface, one must overcome the friction at rest, called static friction. Static friction is about 20% greater than kinetic friction. Kinetic friction is the force that must be overcome to keep a surface moving over another surface.

Different materials have different friction coefficients. The friction coefficient, k, between two surfaces is defined as the ratio of the force, F, needed to move one surface over the other to the total force, W, pressing the two surfaces together.

$$k = F/W$$

Some values of friction coefficients for different surfaces are given in the Reference Table 1 at the beginning of this book. The values of friction coefficients depend on the sliding conditions and range from 1.5 to 0.07.

TEACHER'S SECTION
FRICTION AND THE NATURE OF SURFACES

GOALS AND OBJECTIVES:

1. To learn that friction is a force.
2. To learn that the force due to friction opposes the motion of a body.
3. To learn that the force due to friction is measured in newtons.
4. To learn that the force needed to move an object increases with increasing roughness of the surfaces that are in contact.

DISCUSSION BEFORE THE LABORATORY ACTIVITY:

It is very important to stress the need for wearing an apron and safety goggles.

Discuss Newton's first law of motion. An object at rest remains at rest unless a force acts on the object to move it. A body in motion remains in motion in the same direction and at constant speed, provided there is no net force acting upon it to change the motion. Matter resists any change in its state of motion or rest. This property of matter is called inertia.

Discuss force. Force is a push or a pull that changes the motion of an object. In order to stop this change in motion of the object, another opposite and equal force of push or pull should act on the object.

Discuss friction. There is friction between all objects whenever they make contact with one another. Friction is a force. It resists the motion between any two bodies in contact. A body in motion can be slowed down or completely stopped by the force due to friction.

Show examples of friction. Roll a rubber ball on the floor, and watch its straight–line motion. The ball gradually slows down and finally stops moving. Why didn't the ball continue to move at the same speed? Why did it violate the law of inertia? Ask the students why the ball slowed down and why it stopped moving after some time.

There is a net force acting on the ball. This force is not balanced by any other opposing force. The net force acting on the ball is the force due to friction. This force is exerted by the floor. The extent of this force depends on the nature of the surface of the floor. The surface of one object offers resistance to the movement of the surface of another object over it. This resistance of movement is called friction. Air also exerts a force due to friction. However, it is much less than the force due to friction exerted by the floor.

Measure the distance moved by the ball. Ask the students to suggest ways of making the ball move a longer distance by applying only the same initial force for starting the motion.

Why do we have to continuously supply gas to the engine to drive a car? (To continuously overcome the friction caused by the road and the air.) Friction causes problems for machines. It causes machines to overheat and parts to wear out.

Rough surfaces increase friction. However, a property of the contacting materials, called the friction coefficient, does not depend on the roughness of the surfaces. You may slip while walking on a polished floor but you can walk comfortably on the ground because it offers more friction than the polished floor.

Friction also increases when the surfaces are made too smooth. For example, cleaved mica has very smooth surface. The friction of mica surfaces is as great as that of ordinary surfaces.

The force due to friction also depends on the force used to push one surface (of one object) to slide over the other surface (of the second object). The greater the force used to push an object, the greater the force due to friction. Therefore, it is much harder to push a box with a television inside than the empty box.

Different materials have different friction coefficients. The friction coefficient, k, between two surfaces is defined as the ratio of the force, F, needed to move one surface over the other to the total force, W, pressing the two surfaces together.

$$k = F/W$$

Some values of friction coefficients for different surfaces are given in the Reference Table 1 at the beginning of this book. The values of friction coefficients depend on the sliding conditions and range from 1.5 to 0.07. Metals have higher friction than nonmetals. The friction for metals lessens after lubrication. Elastic materials, such as rubber,, have much more friction than other nonmetals. The use of rubber in automobile tires, shoe soles and heels, provides good traction. Graphite and Teflon are used in coatings that require low friction. Skating and skiing make use of the low friction of ice. Our knee joints have very low friction coefficients of the order of 0.02.

TEACHER'S SECTION
FIGHTING FRICTION

GOALS AND OBJECTIVES:

1. To learn that friction is a force.

2. To learn that the force due to friction opposes the motion of a body.

3. To learn that it is easier to move an object from one place to another by rolling it over wooden dowels than by pushing it over the table.

DISCUSSION BEFORE THE LABORATORY ACTIVITY:

It is very important to stress the need for wearing an apron and safety goggles.

Discuss friction. There is friction between all objects whenever they make contact with one another. Friction is a force. It resists the motion between any two bodies in contact. A body in motion can be slowed down or can be completely stopped by the force due to friction.

Why do we have to continuously supply gas to the engine to drive a car? (To continuously overcome the friction caused by the road and the air.) Friction causes problems to machines. It causes them to overheat and their parts to wear out.

Rough surfaces increase friction. You may slip while walking on a polished floor but you can walk comfortably on the ground, because the ground offers more friction than the polished floor.

Different materials have different friction coefficients. The friction coefficient, k, between two surfaces is defined as the ratio of the force, F, needed to move one surface over the other to the total force, W, pressing the two surfaces together.

$$k = F/W$$

Some values of friction coefficients for different surfaces are given in the Reference Table 1 at the beginning of this book. The values of friction coefficients depend on the sliding conditions and has a range from 1.5 to 0.07. Metals have higher friction than nonmetals. The friction for metals lessens after lubrication. Elastic materials, such as rubber, have much more friction than other nonmetals. The use of rubber in automobile tires, shoe soles and heels, provides good traction. Graphite and Teflon are used in coatings that require low friction. Skating and skiing make use of the low friction of ice. Our knee joints have very low friction coefficients of the order of 0.02.

Rolling is much easier than sliding because rolling produces less friction. Ball bearings give friction coefficients in the range of 0.002 to 0.005. Rolling logs and the use of wheels make use of this lowering of friction.

A car moving with a speed of 100 kilometers per hour (60 miles per hour) has a friction coefficient of 0.8 on a dry road; it needs a stopping distance of 46 meters or 151 feet. The same car on a wet road has a friction coefficient of 0.5 and needs a stopping distance of 74 meters or 242 feet. The same car on an icy road has a friction coefficient of 0.10 and needs a stopping distance of 369 meters or 1210 feet. To avoid jerky motion, brakes and clutches must get constant friction.

We require friction coefficient values of 0.2 or more for normal walking. We can still walk satisfactorily with short strides with friction coefficients of the order of 0.1.

TEACHER'S SECTION
REDUCING FRICTION

GOALS AND OBJECTIVES:

1. To learn that friction is a force.
2. To learn that the force due to friction opposes the motion of a body.
3. To find different ways of reducing friction.
4. To find the least amount of force needed to move an object from one place to another.

DISCUSSION BEFORE THE LABORATORY ACTIVITY:

It is very important to stress the need for wearing an apron and safety goggles.

Discuss friction. There is friction between all objects whenever they make contact with one another. Friction is a force. It resists the motion between any two bodies in contact. A body in motion can be slowed down or completely stopped by the force due to friction.

Show examples of friction. Roll a rubber ball on the floor, and watch its straight-line motion. The ball gradually slows down and finally stops moving. Why didn't the ball continue to move at the same speed? Why did it violate the law of inertia? Ask the students why the ball slowed down and why it stopped moving after a while.

There is a net force acting on the ball. This force is not balanced by any other opposing force. The net force acting on the ball is the force due to friction. This force is exerted by the floor. The extent of this force depends on the nature of the surface of the floor. The surface of one object offers resistance to the movement of the surface of another object over it. This resistance of movement is called friction. Air also exerts a force due to friction. However, it is much less than the force due to friction exerted by the floor.

In devices that move continuously, such as electric motors, friction is a problem. It can cause machine parts to wear out and to overheat. In any machine, the work input is greater than the work output because of friction. In tennis tournaments, the balls are changed regularly after several games, because the friction produced by the hard hitting and bouncing makes the balls wear out easily.

Various techniques can be used for reducing friction. For example, friction in machines can be reduced by using grease and oil, which reduces the wearing out of the machine parts.

TEACHER'S SECTION
THE GOOD ASPECTS OF FRICTION

GOALS AND OBJECTIVES:

1. To learn that friction is a force.
2. To learn that the force due to friction opposes the motion of a body.
3. To learn that friction is often quite useful.

DISCUSSION BEFORE THE LABORATORY ACTIVITY:

It is very important to stress the need for wearing an apron and safety goggles.

Discuss friction. There is friction between all objects whenever they make contact with one another. Friction is a force. It resists the motion between any two bodies in contact. A body in motion can be slowed down or completely stopped by the force due to friction.

Why do we have to continuously supply gas to the engine to drive a car? (To continuously overcome the friction caused by the road and the air.) Friction causes problems for machines. It causes machines to overheat and its parts to wear out.

Friction is essential for controlling motion. Rough surfaces increase friction. You can walk or run comfortably on the ground because of friction between your feet and the ground. You can slip while walking or running on a polished floor or an oil-slick road because of too little friction. A moving car slows down or stops on applying the brake because of friction between the brakes and the wheel. Snow tires with good treads or steel belts make driving a lot safer on a snow-covered road because of more friction. To grip an object, you need friction.

Different materials have different friction coefficients. The friction coefficient, k, between two surfaces is defined as the ratio of the force, F, needed to move one surface over the other to the total force, W, pressing the two surfaces together.

$$k = F/W$$

Some values of friction coefficients for different surfaces are given in the Reference Table 1 at the beginning of this book. The values of friction coefficients depend on the sliding conditions and range from 1.5 to 0.07.

Metals have higher friction than nonmetals. The friction for metals lessens after lubrication. Elastic materials such as rubber have much more friction than other nonmetals. The use of rubber in automobile tires and shoe soles and heels, provides good traction. Graphite and Teflon are used in coatings that require low friction. Skating and skiing make use of the low friction of ice. Our knee joints have very low friction coefficients of the order of 0.02.

Rolling is much easier than sliding because rolling produces less friction. Ball bearings give friction coefficients in the range of 0.002 to 0.005. Rolling logs and the use of wheels make use of this lowering of friction.

A car moving at a speed of 100 kilometers per hour (60 miles per hour) has a friction coefficient of 0.8 on a dry road. The stopping distance of this car is 46 meters or 151 feet. The same car on a wet road has a friction coefficient of 0.5 and needs a stopping distance of 74 meters or 242 feet. The same car on an icy road has a friction coefficient of 0.10 and needs a stopping distance of 369 meters, or 1,210 feet. To avoid jerky motion, one must get constant friction in brakes and clutches.

We require friction coefficient values of 0.2 or more for normal walking. We can still walk satisfactorily with short strides with friction coefficients of the order of 0.1.

Bowed stringed instruments produce music using frictional oscillations. Here the laws of friction are not strictly obeyed.

TEACHER'S SECTION
CHECK YOUR HUGGING POWER

GOALS AND OBJECTIVES:

1. To learn that gases can be compressed by applying pressure.
2. To learn that when a gas is compressed, it occupies less volume.
3. To learn that an object floats in water when its density is less than that of water.
4. To learn that the density of an object containing water and air can be increased by squeezing the air and allowing more water to occupy that space.

DISCUSSION BEFORE THE LABORATORY ACTIVITY:

It is very important to stress the need for wearing an apron and safety goggles.

Discuss the various properties of air, and discuss ways of determining its properties.

Discuss the properties of the three phases of matter: solid, liquid, and gas.

Challenge the students to push a balloon that is slightly bigger than the mouth of the bottle into a bottle. Discuss the reasons why this is difficult to do.

Discuss the meaning of air pressure. The pressure of a gas is caused by collisions of molecules with the walls of the container. The pressure increases with an increase in the number of molecules and an increase in temperature. The normal air pressure at sea level is one atmosphere (atm). A gas can be compressed more easily than a liquid or a solid.

Discuss the densities of solids, liquids, and gases. An object will float in water when its density is less than that of water.

A test tube containing some water when inverted into a bottle completely filled with water will float because the density of test tube – air – water combination is less than the density of water. This test tube can sink to the bottom or to any level by compressing the air inside the test tube and thus, forcing more water into the test tube. This compression of the air can be achieved by applying pressure on the water after closing the bottle.

TEACHER'S SECTION
STATIC ELECTRICITY

GOALS AND OBJECTIVES:

1. To produce static electricity.
2. To learn that in static electricity, the electrical charges do not move.
3. To learn that friction between two nonmetallic surfaces produces static electricity.

DISCUSSION BEFORE THE LABORATORY ACTIVITY:

It is very important to stress the need for wearing an apron and safety goggles.

In static electricity, electrical charges are accumulated. These charges do not move. In current electricity, the charges flow rapidly.

Several methods can be used to produce electric charges on an object. Rubbing two nonmetallic surfaces is the most common way of producing static electricity.

The Greek scientist, Thales of Miletus, produced static electricity as early as 600 B.C. He found that amber, when rubbed with animal fur, attracted dust particles. In the sixteenth century, the English scientist, William Gilbert, discovered this property in various materials.

Benjamin Franklin, in 1752, proved that lightning also is associated with static electricity.

Static electricity is also known as triboelectricity because of its close association with friction between two surfaces (*tribo* is from a Greek word meaning to rub). Many workers have arranged different materials in a triboelectric series. Some examples are given in the Reference Table 2 at the beginning of this book. The material which is on the top of the list gets a positive charge when rubbed against the others. The one at the bottom gets a negative charge when rubbed against the others. In 1757, J. C. Wilcke showed that rubbing any two of the materials on the list will produce positive and negative charges, the material with the positive charge being the one higher on the list.

It is extremely difficult to predict the extent of charging that will be produced. When the two materials in the series are close to each other, it is also difficult to predict the sign of the charges on the materials.

Introduce the concept of atoms. All atoms are neutral. An atom has a nucleus at the center. The center contains protons and neutrons. Protons have an assigned positive charge (+1), and neutrons have no charge. In an atom, electrons are moving around the nucleus. Electrons have an assigned negative charge (−1). Electrons, protons, and neutrons are called fundamental particles. In a neutral atom, the number of protons and the number of electrons are the same. Different elements have different numbers of protons and hence different numbers of electrons. The number of neutrons also varies in different elements.

Some atoms lose electrons easily. Rubbing, or friction between two materials helps one of the materials to lose some electrons. This materials accumulates positive charges. The material that receives electrons accumulates negative charges.

When you scuff your feet on a carpet for several seconds, you collect electrons from the rug and get charged with static electricity. When you touch another person, both of you will get an electrical shock. Some of the extra electrons are transferred quickly into the other person. This process, which takes place very fast is called discharging. If it takes place in the dark, you may be able to see a spark. You can see a spark and receive a shock even when you touch a metallic door knob. People often hear a crackling noise when they separate clothes taken from a dryer. This is caused by electrical discharge.

TEACHER'S SECTION
TWO KINDS OF STATIC ELECTRICITY

GOALS AND OBJECTIVES:

1. To produce static electricity.
2. To learn that in static electricity, the electrical charges do not move.
3. To learn that friction between two nonmetallic surfaces produces static electricity.
4. To learn that there are two kinds of static electricity: positively charged and negatively charged.
5. To learn that positive and negative charges attract each other.
6. To learn that the charge produced on an object depends on the object and the material used for rubbing the object.
7. To learn that the charge produced on a rubber rod by rubbing with fur is different from the charge produced on a glass rod by rubbing with silk.

DISCUSSION BEFORE THE LABORATORY ACTIVITY:

It is very important to stress the need for wearing an apron and safety goggles.

In static electricity, electrical charges are accumulated. These charges are at rest. In current electricity, the charges flow rapidly.

Several methods can be used to produce electric charges on an object. Rubbing two nonmetallic surfaces is the most common way of producing static electricity.

The Greek scientist, Thales of Miletus, had produced static electricity as early as 600 B.C. He found that amber, when rubbed with animal fur, attracted dust particles. In the sixteenth century, the English scientist, William Gilbert, discovered this property in many other materials.

Benjamin Franklin, in 1752, proved that lightning also is associated with static electricity.

Static electricity is also known as triboelectricity because of its close association with friction between two surfaces(*tribo* is from a Greek word meaning to rub). Many workers have arranged different materials in a triboelectric series. Some examples are given in the Reference Table 2 at the beginning of this book. The first material on the list gets a positive charge when rubbed against the others. The one at the bottom gets a negative charge when rubbed against the others. In 1757, J. C. Wilcke showed that rubbing any two of the materials on the list will produce positive and negative charges, the material with the positive charge being the one higher on the list.

It is extremely difficult to predict the extent of charging that will be produced. When the two materials in the series are close to each other, it is also difficult to predict the sign of the charges on the materials.

Matter is made up of atoms of different elements. An atom has a nucleus, containing positively charged protons and neutrons with no charge. Negatively charged electrons revolve around the nucleus. The number of protons and electrons in an atom are equal. It is possible to transfer electrons from one object to another object by rubbing. It is not possible to transfer protons from one object to another. The material that loses the electrons gets the positive charge, and the material that gets the electrons becomes negatively charged. Rubbing two objects gives them the ability to attract tiny bits of matter.

Discuss with students the fact that an electrically charged object attracts all uncharged objects. It will also attract all objects with an unlike charge. On the other hand, two objects with like charges repel each other. Thus repulsion is a sure way of identifying an object with a charge.

When a rubber rod is rubbed with fur, it acquires a negative charge, and the fur gets an equal positive charge. When a glass rod is rubbed with silk, it gets a positive charge and the silk acquires an equal negative charge. The two rubber rods will repel each other. The two glass rods also will repel each other. The rubber rod and the glass rod, on the other hand, will attract each other.

Rubbing produces equal and opposite charges on the two objects. Combing your hair produces negative charges on the comb and an equal number of positive charges on the hair. Rubbing a balloon with a sweater produces negative charges on the balloon and an equal number of positive charges on the sweater.

TEACHER'S SECTION
ATTRACTION AND REPULSION BETWEEN ELECTRIC CHARGES

GOALS AND OBJECTIVES:

1. To produce static electricity, or electricity at rest.
2. To learn that there are two kinds of static electricity, positively charged and negatively charged.
3. To learn that positive and negative charges attract each other.
4. To learn that two positive charges repel each other.
5. To learn that two negative charges repel each other.

DISCUSSION BEFORE THE LABORATORY ACTIVITY:

It is very important to stress the need for wearing an apron and safety goggles.

In static electricity, electrical charges are accumulated. These charges are at rest. In current electricity, the charges flow rapidly.

Several methods can be used to produce electric charges on an object. Rubbing two nonmetallic surfaces together is the most common way of producing static electricity. It is possible to electrify any object by rubbing it with another object. Combing hair with a plastic comb and walking across a nylon carpet with rubber shoes are examples of this electrification.

The Greek scientist, Thales of Miletus, had produced static electricity as early as 600 B.C. He found that amber, when rubbed with animal fur, attracted light objects. In the sixteenth century, the English scientist, William Gilbert, discovered this property in many other materials.

Benjamin Franklin, in 1752, proved that lightning is also associated with static electricity.

Static electricity also is known as triboelectricity because of its close association with friction between two surfaces(*tribo* is from a Greek word meaning to rub). Many workers have arranged different materials in a triboelectric series. Some examples are given in the Reference Table 2 at the beginning of this book. The first material on the list gets a positive charge when rubbed against the others. The one at the bottom of the list gets a negative charge when rubbed against the others. In 1757, J. C. Wilcke showed that rubbing any two of the materials on the list will produce positive and negative charges, the material with the positive charge being the one higher on the list.

It is extremely difficult to predict the extent of charging that will be produced. When the two materials in the series are close to each other, it is also difficult to predict the sign of the charges on the materials.

The materials at the two ends of the triboelectric series can acquire only one type of charge, either positive (the top end of the series) or negative (the bottom end of the series). However, materials at the middle of the series can acquire either a positive charge or a negative charge by selecting another suitable material for rubbing.

Two suspended pith balls or two cork balls, when touched by a plastic rod that has been charged by rubbing it with fur, acquire the same negative charge of the plastic rod. The pith balls or the cork balls and the plastic rod repel each other. The two pith balls or the two cork balls also repel each other. Similarly, two balls of pith or cork, when touched by a glass rod charged by rubbing it with silk, acquire the same positive charge as that of the glass rod. The pith or the cork balls and the glass rod then repel each other. The two balls of pith or cork also repel each other. When the balls of pith or cork charged by the plastic rod are brought near the balls of pith or cork charged by the glass rod, they attract each other. This shows that the pith balls or the cork balls can acquire either a positive charge or a negative charge depending on the material used for charging them.

Matter is made up of atoms of different elements. An atom has a nucleus, containing positively charged protons and neutrons with no charge. Negatively charged electrons revolve around the nucleus. An atom has an equal number of protons and electrons. It is possible to transfer electrons from one object to another object by rubbing. It is not possible to transfer protons from one object to another. The material that loses the electrons gets the positive charge, and the material that gets the electrons becomes negatively charged. Rubbing two objects together gives them the ability to attract tiny bits of matter.

Discuss with students the fact that an electrically charged object attracts all uncharged objects. It also attracts all objects with an unlike charge. On the other hand, two objects with like charges repel each other. Thus repulsion is a sure way of identifying an object with a charge.

When a great difference in charges between two objects is built up, a spark will be produced when the negative charges jump to the positive charges because of attraction. Walking on a carpet with rubber shoes helps to build up negative charges on a person. When the person touches a metal, such as a door knob, negative charges jump through the dry air as a spark to the door knob. Lightning between clouds or between a cloud and the ground is a giant spark produced when the negative charges on one cloud jump to the positive charges on another cloud or to the positive charges on the ground. These sparks are caused by attraction between opposite charges.

When a balloon is rubbed with wool, the balloon acquires a negative charge, and the wool gets an equal positive charge. Two negatively charged balloons will repel each other.

Rubbing produces equal and opposite charges on the balloon and the wool.

TEACHER'S SECTION
ELECTRICAL FORCE AND GRAVITATIONAL FORCE

GOALS AND OBJECTIVES:

1. To produce static electricity, or electricity at rest.
2. To learn that charged objects attract uncharged objects.
3. To learn that the electrical force of attraction is different from the gravitational force of attraction.
4. To learn that the electrical force of attraction between two objects may be greater than the gravitational force of attraction between the charged object and the earth.

DISCUSSION BEFORE THE LABORATORY ACTIVITY:

It is very important to stress the need for wearing an apron and safety goggles.

Rubbing is the most common way of producing static electrical charges on an object. It is possible to electrify any object by rubbing it with another object.

Static electricity also is known as triboelectricity because of its close association with friction between two surfaces (*tribo* is from a Greek word meaning to rub). Several workers have arranged different materials in a triboelectric series. Some examples are given in the Reference Table 2 at the beginning of this book. The first material on the list gets a positive charge when rubbed against the others. The one at the bottom of the list gets a negative charge when rubbed against the others. In 1757, J. C. Wilcke showed that rubbing any two of the materials on the list will produce positive and negative charges, the material with the positive charge being the one higher on the list.

It is extremely difficult to predict the extent of charging that will be produced. When the two materials in the series are close to each other, it is also difficult to predict the sign of the charges on the materials.

The materials at the two ends of the triboelectric series can acquire only one type of charge, either positive (the top end of the series) or negative (the bottom end of the series). However, materials at the middle of the series can acquire either a positive charge or a negative charge by selecting another suitable material for rubbing.

Discuss with students the fact that an electrically charged object attracts any uncharged object. It will also attract any object with an unlike charge. On the other hand, two objects with like charges repel each other. Thus repulsion is a sure way of identifying an object with a charge.

When a balloon is rubbed with wool, the balloon acquires a negative charge, and the wool gets an equal positive charge. The negatively charged balloon can cling to a wall or to the bottom of a wooden table by electrostatic attraction. In this activity, students will find out that the electrostatic force of attraction between the charged balloon and the bottom of a table can support some mass of matter against gravity.

When an object is thrown up in the air, it falls down because of the gravitational force of attraction between the object and the earth.

The gravitational force of attraction depends on the mass of the object, the mass of the earth, and the distance between the object and the center of the earth. There is gravitational force of attraction between any two objects. Since the masses of different objects are very small compared with the mass of the earth, the gravitational force of attraction between the two objects is also very small compared with the gravitational force of attraction between the object and the earth.

If the electrostatic force of attraction between a charged object and a wall is greater than the gravitational force of attraction between the object and the earth, the object will cling to the wall. When the mass of the charged object is increased more and more, the gravitational force also will increase continuously. When the gravitational force exceeds the electrostatic force, the object cannot be suspended and will fall down.

TEACHER'S SECTION
SUSPENDING COINS USING STATIC ELECTRICITY

GOALS AND OBJECTIVES:

1. To produce static electricity, electricity at rest.

2. To learn that charged objects attract uncharged objects.

3. To learn that the electrical force of attraction is different from the gravitational force of attraction.

4. To learn that the electrical force of attraction between two objects may be greater than the gravitational force of attraction between the charged object and the earth.

5. To find the mass of matter whose gravitational force of attraction with the earth will balance the electrostatic force of attraction between the charged object and the ceiling (bottom of a table).

DISCUSSION BEFORE THE LABORATORY ACTIVITY:

It is very important to stress the need for wearing an apron and safety goggles.

Several methods can be used to produce static electric charges on an object. Rubbing two nonmetallic surfaces together is the most common way of producing static electricity. It is possible to electrify any object by rubbing it with another object.

Static electricity is also known as triboelectricity because of its close association with friction between two surfaces (*tribo* is from a Greek word meaning to rub). Several workers have arranged different materials in a triboelectric series. Some examples are given in the Reference Table 2 at the beginning of this book. The first material on the list gets a positive charge when rubbed against the others. The one at the bottom of the list gets a negative charge when rubbed against the others. In 1757, J. C. Wilcke showed that rubbing any two of the materials on the list will produce positive and negative charges, the material with the positive charge being the one higher on the list.

It is extremely difficult to predict the extent of charging that will be produced. When the two materials in the series are close to each other, it is also difficult to predict the sign of the charges on the materials.

The materials at the two ends of the triboelectric series can acquire only one type of charge, either positive (the top end of the series) or negative (the bottom end of the series). However, materials at the middle of the series can acquire either a positive charge or a negative charge by selecting another suitable material for rubbing.

Discuss with students the fact that an electrically charged object attracts any uncharged object. It will also attract any object with an unlike charge. On the other hand, two objects with like charges repel each other. Thus repulsion is a sure way of identifying an object with a charge.

When a balloon is rubbed with wool, the balloon acquires a negative charge, and the wool gets an equal positive charge. The negatively charged balloon can cling to a wall or to the bottom of a wooden table by electrostatic attraction. In this activity, students will find out that the electrostatic force of attraction between the charged balloon and the bottom of a table can support some mass of matter against gravity.

When an object is thrown up in the air, it falls down because of the gravitational force of attraction between the object and the earth.

The gravitational force of attraction depends on the mass of the object, the mass of the earth, and the distance between the object and the center of the earth. There is gravitational force of attraction between any two objects. Since the masses of different objects are very small compared with the mass of the earth, the gravitational force of attraction between the two objects is also very small compared with the gravitational force of attraction between the object and the earth.

If the electrostatic force of attraction between a charged object and a wall is greater than the gravitational force of attraction between the object and the earth, the object will cling to the wall. When the mass of the charged object is increased more and more, the gravitational force also will increase continuously. When the gravitational force exceeds the electrostatic force, the object cannot be suspended and will fall down.

TEACHER'S SECTION
SUSPENDING A CHARGED BALLOON FROM DIFFERENT SURFACES

GOALS AND OBJECTIVES:

1. To produce static electricity, or electricity at rest.
2. To learn that charged objects attract uncharged objects.
3. To learn that the electrical force of attraction is different from the gravitational force of attraction.
4. To learn that the electrical force of attraction between two objects may be greater than the gravitational force of attraction between the charged object and the earth.
5. To find which surface has maximum electrostatic attraction between that surface and a balloon charged by wool.

DISCUSSION BEFORE THE LABORATORY ACTIVITY:

It is very important to stress the need for wearing an apron and safety goggles.

Rubbing is the most common way of producing static electrical charges on an object. It is possible to electrify any object by rubbing it with another object.

Static electricity is also known as triboelectricity because of its close association with friction between two surfaces (*tribo* is from a Greek word meaning to rub). Several workers have arranged different materials in a triboelectric series. Some examples are given in the Reference Table 2 at the beginning of this book. The first material on the list gets a positive charge when rubbed against the others. The one at the bottom of the list gets a negative charge when rubbed against the others. In 1757, J. C. Wilcke showed that rubbing any two of the materials on the list will produce positive and negative charges, the material with the positive charge being the one higher on the list.

It is extremely difficult to predict the extent of charging that will be produced. When the two materials in the series are close to each other, it is also difficult to predict the sign of the charges on the materials.

Discuss with students the fact that an electrically charged object attracts any uncharged object. It will also attract any object with an unlike charge. On the other hand, two objects with like charges repel each other. Thus repulsion is a sure way of identifying an object with a charge.

When a balloon is rubbed with wool, the balloon acquires a negative charge, and the wool gets an equal positive charge. The negatively charged balloon can cling to a wall or to the bottom of a wooden table by electrostatic attraction.

In this activity, students will find out that the electrostatic force of attraction between the charged balloon and the bottom of a table can support some mass of matter against gravity.

When an object is thrown up in the air, it falls down because of the gravitational force of attraction between the object and the earth.

The gravitational force of attraction depends on the mass of the object, the mass of the earth, and the distance between the object and the center of the earth. There is gravitational force of attraction between any two objects. Since the masses of different objects are very small compared with the mass of the earth, the gravitational force of attraction between the two objects is also very small compared with the gravitational force of attraction between the object and the earth.

If the electrostatic force of attraction between a charged object and a wall is greater than the gravitational force of attraction between the object and the earth, the object will cling to the wall. When the mass of the charged object is increased more and more, the gravitational force also will increase continuously. When the gravitational force exceeds the electrostatic force, the object cannot be suspended and will fall down.

TEACHER'S SECTION
STYROFOAM CUP CONTEST

GOALS AND OBJECTIVES:

1. To produce static electricity, electricity at rest.
2. To learn that charged objects attract uncharged objects.
3. To learn that the electrical force of attraction is different from the gravitational force of attraction.
4. To learn that the electrical force of attraction between two objects may be greater than the gravitational force of attraction between the charged object and the earth.
5. To find which material — wool, silk, cotton or nylon — can produce the maximum static electric charge on a Styrofoam cup.

DISCUSSION BEFORE THE LABORATORY ACTIVITY:

It is very important to stress the need for wearing an apron and safety goggles.

Rubbing is the most common way of producing static electrical charges on an object. It is possible to electrify any object by rubbing it with another object. A rubber balloon was charged with static electricity in several previous activities. In this activity, a Styrofoam cup will be charged by rubbing it with different materials and the extent of charging will be investigated by competing with gravitational force of attraction.

Static electricity is also known as triboelectricity because of its close association with friction between two surfaces(*tribo* is from a Greek word meaning to rub). Several workers have arranged different materials in a triboelectric series. Some examples are given in the Reference Table 2 at the beginning of this book. The first material on the list gets a positive charge when rubbed against the others. The one at the bottom of the list gets a negative charge when rubbed against the others. In 1757, J. C. Wilcke showed that rubbing any two of the materials on the list will produce positive and negative charges, the material with the positive charge being the one higher on the list.

It is extremely difficult to predict the extent of charging that will be produced. When the two materials in the series are close to each other, it is also difficult to predict the sign of the charges on the materials.

Discuss with students the fact that an electrically charged object attracts any uncharged object. In this activity students will find out whether the attraction between the charged Styrofoam cup and the uncharged wall is the same or different when different materials are used for charging. If the electrostatic force of attraction between the charged Styrofoam cup and the wall can support the same amount of matter against gravity, no matter what the material is used for rubbing the Styrofoam cup, then it will indicate that the extent of charging is the same in each case. If the amount of matter that can be suspended is different, then the extent of charging the Styrofoam cup depends on the material used for rubbing.

When an object is thrown up in the air, it falls down because of the gravitational force of attraction between the object and the earth.

The gravitational force of attraction depends on the mass of the object, the mass of the earth, and the distance between the object and the center of the earth. There is gravitational force of attraction between any two objects. Since the masses of different objects are very small compared with the mass of the earth, the gravitational force of attraction between the two objects is also very small compared with the gravitational force of attraction between the object and the earth.

If the electrostatic force of attraction between a charged object and a wall is greater than the gravitational force of attraction between the object and the earth, the object will cling to the wall. When the mass of the charged object is increased more and more, the gravitational force also will increase continuously. When the gravitational force exceeds the electrostatic force, the object cannot be suspended and will fall down.

TEACHER'S SECTION
IDENTIFYING WATER AND OIL USING STATIC ELECTRICITY

GOALS AND OBJECTIVES:

1. To produce static electricity on a balloon and on a Styrofoam cup by rubbing them against wool.

2. To learn that a stream of water is deflected toward an object with a charge.

3. To learn that a water molecule has ends with positive and negative charges, and these charged ends can easily attract other charged objects.

4. To learn that a stream of baby oil is not deflected towards an object charged with static electricity.

5. To learn that a molecule of baby oil and a molecule of water behave differently towards charged objects.

DISCUSSION BEFORE THE LABORATORY ACTIVITY:

It is very important to stress the need for wearing an apron and safety goggles.

In static electricity, electrical charges are accumulated. These charges do not move. In current electricity, the charges flow rapidly.

Several methods can be used to produce electric charges on an object. Rubbing two nonmetallic surfaces together is the most common way of producing static electricity.

Static electricity is also known as triboelectricity because of its close association with friction between two surfaces(*tribo* is from a Greek word meaning to rub). Many workers have arranged different materials in a triboelectric series. Some examples are given in the Reference Table 2 at the beginning of this book. The first material on the list gets a positive charge when rubbed against the others. The one at the bottom of the list gets a negative charge when rubbed against the others. In 1757, J. C. Wilcke showed that rubbing any two of the materials on the list will produce positive and negative charges, the material with the positive charge being the one higher on the list.

It is extremely difficult to predict the extent of charging that will be produced. When the two materials in the series are close to each other, it is also difficult to predict the sign of the charges on the materials.

Certain molecules, such as water and ammonia have ends with positive and negative charges. A stream of these molecules will be deflected toward an electric charge, either positive or negative, because of strong attraction between unlike charges. A charged object attracts all uncharged objects and also unlike charges. However, the attraction between the unlike charges is much stronger than the attraction between the charged object and the uncharged object. This is evident from the fact that a charged object is unable to deflect a stream of baby oil. A molecule of baby oil is long, with practically no charged ends like water. The same is true of hexane (similar to gasoline).

348 Fundamentals of Science Activity Series

TEACHER'S SECTION
ELECTRICAL CIRCUITS

GOALS AND OBJECTIVES:

1. To learn that electrons flow in current electricity.
2. To learn that metals are good conductors of electricity.
3. To learn that an electrical circuit describes the path of electron flow.
4. To learn that there is no electron flow in an open circuit.
5. To learn that electrons flow through a closed circuit.
6. To learn that a battery is a source of electrons.

DISCUSSION BEFORE THE LABORATORY ACTIVITY:

It is very important to stress the need for wearing an apron and safety goggles.

Static electricity is electricity at rest. In static electricity, electrical charges are accumulated. Static electricity is most commonly produced by rubbing two nonmetallic surfaces so that there is a transfer of electrons from one surface to the other. The position of a material in the triboelectric series determines whether it will lose electrons. The material at the top of the series transfers its electrons to all the materials below it in the series.

A current is a flow of electric charges. Current electricity is the flow of electrons. All dry cell batteries and lead–acid batteries used in cars provide a source of electrons. Chemical reactions in these batteries provide the electrons.

There are two kinds of electric currents: direct current and alternating current. In direct current (DC), the electrons always travel in the same direction. (See illustration 25–1.) Batteries or dry cells produce direct currents from chemical energy. In alternating currents (AC), the electrons travel first in one direction, then they move in the opposite direction. Thus the electrons travel back

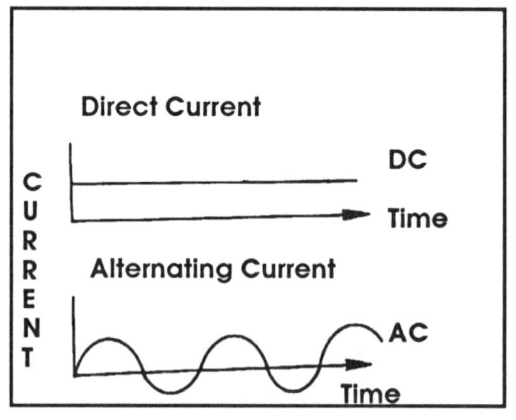

and forth. Alternating current, produced by electric power companies and used in our homes, offices, and industries, is cheaper to produce. It is also easier to transmit over long distances.

Discuss what are the things needed to have an electric current or flow of electricity. We need a large supply of electrons that are free to move. We need a source of energy to push these free electrons. We need a material through which electrons can move with ease. We also need an electrical circuit, or pathway.

Discuss the meaning of electrical circuits, open circuits, and closed circuits. Electrical circuits provide the path for the flow of electrons. The path for the flow of electrons is provided by an electrical conductor, such as a copper wire. Electrons flow from the negative terminal of the battery through the circuit and ends at the positive terminal of the battery. Electricity will flow through a closed circuit, because the electrical current can return to its source. A closed circuit is a complete circuit. Electricity cannot flow through an open circuit, because the electrons coming from the negative terminal cannot reach the positive terminal. An open circuit is not a complete circuit.

The force needed to move the electrons through the circuit is the voltage. The unit of this force is volts. More force is needed to move the electrons through the wires in an electric heater or toaster, because these wires have high resistance.

A switch is used to open or close a circuit so that one can control the flow of electricity.

Electricians use a form of shorthand to represent electrical circuits. Brief shorthand representations of electrical circuits are given in the Reference Table 3 at the beginning of this book. Introduce students to this way of drawing electrical circuits.

In this activity, two dry "D" cells of 1.5 volts are used. You can get two "D" cell connectors from electronic shops. These connectors provide color-coded connections. This eliminates the need for taping the cells in series. Connecting the negative terminal of a cell to the positive terminal of a second cell, the negative terminal of the second cell to the positive terminal of the third cell, and so on will give a series connection.

A 3 volts light bulb is needed for this. The length of the bell wire is not critical.

TEACHER'S SECTION
ELECTRICAL CONDUCTORS AND INSULATORS

GOALS AND OBJECTIVES:

1. To learn that in current electricity, electrons flow.
2. To learn that metals are good conductors of electricity.
3. To learn that nonmetals are poor conductors of electricity.
4. To learn that poor conductors of electricity are called insulators.
5. To learn that a battery is a source of electrons.

DISCUSSION BEFORE THE LABORATORY ACTIVITY:

It is very important to stress the need for wearing an apron and safety goggles.

Current electricity is the flow of electrons. All dry cell batteries and lead–acid batteries used in cars provide a source of electrons. Chemical reactions in these batteries create the electrons.

Discuss the meaning of electrical circuits, open circuits, and closed circuits. Electrical circuits provide the path for the flow of electrons. Electrons flow from the negative terminal of the battery through the circuit and end at the positive terminal of the battery. Electricity will flow through a closed circuit because the electrical current can return to its source. Electricity cannot flow through an open circuit, because the electrons coming from the negative terminal cannot reach the positive terminal.

Metals such as copper and aluminum are good conductors of electricity. Copper wires are used for the electrical wiring in homes. These copper wires are insulated with plastic because plastic does not conduct electricity. (See illustration 26–1.) A poor conductor such as plastic is a good insulator, because it allows you to touch the insulated wire even when the electricity is flowing through the wire. Insulated copper wires of different thicknesses are used,

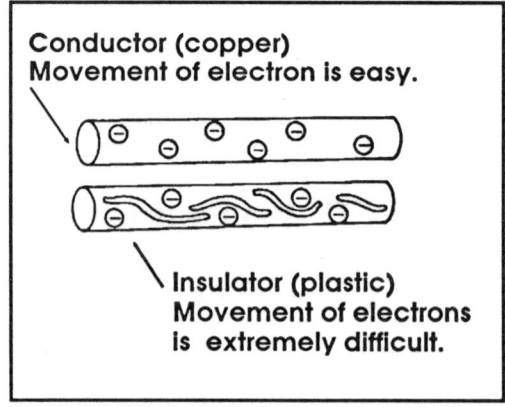

Conductor (copper)
Movement of electron is easy.

Insulator (plastic)
Movement of electrons is extremely difficult.

depending on the amount of current expected to flow through them. Thicker wires can carry more electricity without getting hot.

A good conductor will release its outer electrons easily. These outer electrons are part of the circuit.

Conductors such as copper and aluminum lose a large amount of energy in the form of heat. Superconductors carry electric current without losing energy. However, superconductivity is exhibited only by some metals at very low temperatures. In the past cooling to such low temperatures required expensive liquid helium. More recently, scientists have discovered ceramic superconductors that will work at temperatures that can be achieved by cooling with less expensive liquid nitrogen.

In this activity, two dry 1.5 volts "D" cells are used. You can get two "D" cell connectors from electronic shops. These connectors provide color-coded connections, eliminating the need for taping the cells in a series. Connecting the negative terminal of a cell to the positive terminal of a second cell, connecting the negative terminal of the second cell to the positive terminal of the third cell, and so on, will give a series connection.

A 3 volts light bulb is needed for this activity. The length of the bell wire is not critical.

TEACHER'S SECTION
ELECTRICAL RESISTANCE OF DIFFERENT METALS

GOALS AND OBJECTIVES:

1. To learn that in current electricity, electrons flow.
2. To learn that metals are good conductors of electricity.
3. To learn that different metals conduct electricity differently because of differences in their resistance.
4. To learn that metallic alloys such as nichrome offer much resistance to the flow of electricity.
5. To learn that the resistance of a wire increases with the increasing length of the wire.
6. To learn that the unit for measuring electric current is called an ampere.
7. To learn that the brightness of a light bulb decreases with increasing resistance in the electrical circuit.
8. To learn that the unit for measuring resistance of a wire is called an ohm.

DISCUSSION BEFORE THE LABORATORY ACTIVITY:

It is very important to stress the need for wearing an apron and safety goggles.

Current electricity is the flow of electrons. All dry cell batteries and lead–acid batteries used in cars provide a source of electrons. Chemical reactions in these batteries create the electrons.

Discuss the meaning of electrical circuits, open circuits, and closed circuits. Electrical circuits provide the path for the flow of electrons. Electrons flow from the negative terminal of the battery through the circuit and end at the positive terminal of the battery.

Metals such as copper and aluminum are conductors of electricity. Copper wires are used for the electrical wiring in homes. These copper wires are insulated with plastic, because plastic does not conduct electricity. This allows you to touch the insulated wire even when the electricity is flowing through the wire. Insulated copper wires of different thicknesses are used, depending on the amount of current expected to flow through them. Thicker wires can carry more electricity without getting hot.

The ease with which electrons flow through a metal or metallic alloy depends on the nature of the metal or alloy. The wires used in electrical wiring conduct electricity with ease. The wires used in light bulbs, electrical hot plates, toaster ovens, and space heating offer a great deal of resistance for the flow of electricity, and they dissipate electrical energy in the form of heat.

The approximate resistances of 100 feet of wire (American Wire Gauge Number 30) at 20°Celsius are 10.32 ohms for standard annealed copper, 16.9 ohms for aluminum, 675 ohms for nichrome, and 70.6 for steel piano wire. The resistances of 100 feet of wire (American Wire Gauge No. 24) at 20°Celsius are 2.567 ohms for standard annealed copper, 4.21 ohms for aluminum, 167.1 ohms for nichrome, and 17.6 ohms for steel piano wire. The resistance of steel piano wire (AWG No. 32) is 112 ohms for 100 feet. Steel piano wires of AWG No. 24 and 31 and other metal wires of Gauge Numbers 24 or 30, useful for this activity, are available from scientific supply companies, such as Sargent–Welch.

The resistance of a 1 centimeter cube of a substance to the flow of electricity is known as the specific resistance or resistivity. The current flow is perpendicular to two parallel faces of the cube. This resistivity is characteristic of different substances. It is defined by the expression:

Resistance = resistivity x length/cross sectional area

where the resistance of a uniform conductor is measured in a unit called ohms: the length of the conductor is measured in centimeters; the cross–sectional area of the conductor is measured in square centimeters; and the resistivity of the conductor is measured in a unit called ohm centimeters.

At 20° Celsius, the resistivity values of aluminum, copper (annealed), nichrome, and steel (E. B. B) wires are 0.000002824, 0.0000017241, 0.000100, and 0.0000104 ohm.centimeters respectively. The resistivity of steel depends on the type of steel. These values indicate that copper has the least resistance and nichrome has the most resistance under comparable conditions.

Discuss an ammeter. An ammeter is a devise used for measuring electric currents flowing through a circuit. An ammeter is always connected in series. A series circuit allows only one path for the electron flow. The devise should not be connected unless there is a resistor such as a light bulb in the circuit.

In this activity, two dry 1.5 volts "D" cells are used. You can get two "D" cell connectors from electronic shops. These connectors provide color-coded connections, eliminating the need for taping the cells in a series. Connecting the negative terminal of a cell to the positive terminal of a second cell, connecting the negative terminal of the second cell to the positive terminal of the third cell, and so on will give a series connection.

A 3 volts light bulb is needed for this activity. The length of the bell wire is not critical.

TEACHER'S SECTION
ELECTRICAL RESISTANCE AND LENGTH OF THE METAL WIRE

GOALS AND OBJECTIVES:

1. To learn that in current electricity, electrons flow.

2. To learn that cooper is a good conductor of electricity.

3. To learn that the resistance of a wire increases with increasing length of the wire.

4. To learn that the unit for measuring electric current is called an ampere.

5. To learn that the brightness of a light bulb decreases with increasing resistance in the electrical circuit.

6. To learn that the unit for measuring the resistance of a wire is called an ohm.

DISCUSSION BEFORE THE LABORATORY ACTIVITY:

It is very important to stress the need for wearing an apron and safety goggles.

Discuss the meaning of electrical circuits, open circuits, and closed circuits. Electrical circuits provide the path for the flow of electrons. Electricity will flow only if it can return to its source.

Metals such as copper and aluminum are conductors of electricity. Copper wires are used for the electrical wiring in homes. These copper wires are insulated with plastic, because plastic does not conduct electricity. This allows you to touch the insulated wire even when the electricity is flowing through the wire. Insulated copper wires of different thicknesses are used, depending on the amount of current expected to flow through them. Thicker wires can carry more electricity without getting hot. The resistance of a wire increases with in-

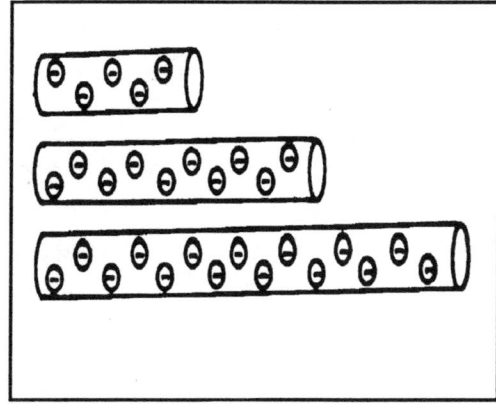

Electrical resistance increases with increasing length of the wire.

creasing length of the wire. (See illustration 28–1.)

The ease with which electrons flow through a metal or metallic alloy depends on the nature of the metal or alloy. The wires used in electrical wiring conduct electricity with ease. The wires used in light bulbs, electrical hot plates, toaster ovens, and space heating offer a great deal of resistance for the flow of electricity, and they dissipate electrical energy in the form of heat.

Resistance is a property of conductors. The current produced by a given potential difference (voltage) is determined by the resistance in the circuit. The resistance of a wire is determined by the nature of the wire, the dimensions of the wire, and the temperature of the wire. The practical unit of resistance is called an ohm. One ohm is the resistance that will produce a current of one ampere when a potential difference of one volt is applied to the circuit. The resistance of 100 feet of standard annealed copper at 20° Celsius is 0.1588 ohm (for American Wire gauge No. 12), 0.2525 ohm (for AWG No. 14), 0.6385 ohm (for AWG No. 18), 10.32 ohms (for AWG No. 30), 41.48 ohms (for AWG No. 36) and 104.9 ohms (for AWG No. 40).

The resistance of a a one centimeter cube of a substance to the flow of electricity is known as the specific resistance or resistivity. The current flow is perpendicular to two parallel faces of the cube. This resistivity is characteristic of different substances. It is defined by the expression:

Resistance = resistivity x length/cross sectional area

where the resistance of a uniform conductor is measured in a unit called ohms: the length of the conductor is measured in centimeters; the cross-sectional area of the conductor is measured in square centimeters; and the resistivity of the conductor is measured in a unit called ohm centimeters.

At 20° Celsius, the resistivity values of annealed copper and hard-drawn copper are 0.0000017241 and 0.000001771 ohm.centimeters.

Discuss an ammeter. An ammeter is a devise used for measuring electric currents flowing through a circuit. An ammeter is always connected in series. A series circuit allows only one path for the electron flow. The devise should not be connected unless there is a resistor such as a light bulb in the circuit.

In this activity, two dry 1.5 volts "D" cells are used. You can get two "D" cell connectors from electronic shops. These connectors provide color-coded connections, eliminating the need for taping the cells in a series. Connecting the negative terminal of a cell to the positive terminal of a second cell, connecting the negative terminal of the second cell to the positive terminal of the third cell, and so on will give a series connection.

A 3 volts light bulb is needed for this activity. The length of the bell wire is not critical.

In this activity, it is better to use wires with the least thickness. Copper wires, American Wire Gauge Numbers 30 or 36, available from scientific supply companies such as Sargent–Welch are suitable for this activity.

TEACHER'S SECTION
ELECTRICAL RESISTANCE AND THICKNESS OF THE METAL WIRE

GOALS AND OBJECTIVES:

1. To learn that copper is a good conductor of electricity.

2. To learn that copper wires of different thicknesses conduct electricity differently because of differences in their resistance.

3. To learn that the resistance of a wire decreases with increasing thickness of the wire.

4. To learn that the unit for measuring electric current is called an ampere.

5. To learn that the brightness of a light bulb decreases with increasing resistance in the electrical circuit.

6. To learn that thick wires should be used for carrying more electricity.

DISCUSSION BEFORE THE LABORATORY ACTIVITY:

It is very important to stress the need for wearing an apron and safety goggles.

Metals such as copper and aluminum are conductors of electricity. Copper wires are used for the electrical wiring in homes. These copper wires are insulated with plastic, because plastic does not conduct electricity. This allows you to touch the insulated wire even when the electricity is flowing through the wire. Insulated copper wires of different thicknesses are used, depending on the amount of current expected to flow through them. Thicker wires can carry more electricity without getting hot. (See illustration 29–1.)

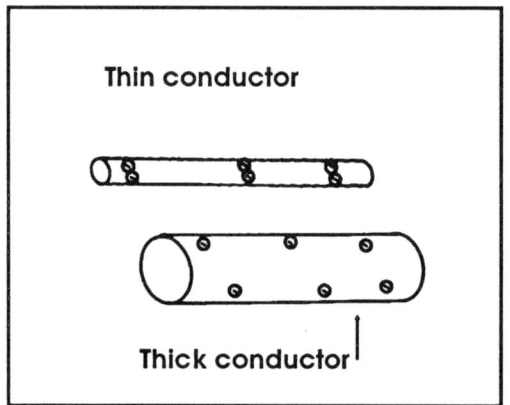

Electrons move more easily through thick conductor than through thin conductor.

The ease with which electrons flow through a metal or metallic alloy depends on the nature of the metal or alloy. The wires used in electrical wiring conduct electricity with ease. The wires used in light bulbs, electrical hot plates, toaster ovens, and space heating offer a great deal of resistance for the flow of electricity, and they dissipate electrical energy in the form of heat.

The approximate resistances of 100 feet of standard annealed copper wire at 20° Celsius are 0.6385 ohm for American Wire Gauge Number 18, 2.567 ohms for AWG No. 24, and 10.32 ohms for AWG No. 30. These wires are suitable for this activity and are available from scientific supply companies such as Sargent–Welch.

Discuss an ammeter. An ammeter is a devise used for measuring electric currents flowing through a circuit. An ammeter is always connected in series. A series circuit allows only one path for the electron flow. The devise should not be connected unless there is a resistor such as a light bulb in the circuit.

In this activity, two dry 1.5 volts "D" cells are used. You can get two "D" cell connectors from electronic shops. These connectors provide color-coded connections, eliminating the need for taping the cells in a series. Connecting the negative terminal of a cell to the positive terminal of a second cell, connecting the negative terminal of the second cell to the positive terminal of the third cell, and so on will give a series connection.

A 3 volts light bulb is needed for this activity. The length of the bell wire is not critical.

360 Fundamentals of Science Activity Series

TEACHER'S SECTION
SERIES CIRCUITS

GOALS AND OBJECTIVES:

1. To learn that in current electricity, electrons flow.
2. To learn that an electrical circuit describes the path of electron flow.
3. To learn that there is no electron flow in an open circuit.
4. To learn that electrons flow through a closed circuit.
5. To learn that in a series circuit, there is only one path for the electricity to flow through.
6. To learn that electricity will not flow, if there is any loose connection or open connection anywhere in a series circuit.

DISCUSSION BEFORE THE LABORATORY ACTIVITY:

It is very important to stress the need for wearing an apron and safety goggles.

Static electricity is electricity at rest. In static electricity, electrical charges are accumulated. Static electricity is produced by rubbing two nonmetallic surfaces so that there is a transfer of electrons from one surface to the other.

Current electricity is the flow of electrons. For the electricity to travel, it must have a path. The complete path through which electricity travels is called a circuit.

There are open circuits and closed circuits. Electrical circuits provide the path for the flow of electrons. Electrons flow from the negative terminal of the battery through the circuit and end at the positive terminal of the battery. Electricity will flow through a closed circuit, because the electrical current can return to its source. Electricity cannot flow through an open circuit, because the electrons coming from the negative terminal cannot reach the positive terminal.

There are series circuits and parallel circuits. In a series circuit, there is only one path for the flow of electricity. If this path has been opened by a loose connection or a loose bulb, electricity will not flow through the circuit. A parallel circuit contains two or more separate paths for the flow of electricity. This will allow the electricity to flow through one path even if the other paths are open.

Resistance is the opposition to the flow of electrons in an electrical circuit. The resistance of a wire depends on the nature and dimensions of the wire.

In a series circuit, the total resistance is the sum of the individual resistances. If the resistance of the connecting wires is not taken into account, then the total resistance is given by the expression:

Total resistance = resistance of light bulb 1 + resistance of light bulb 2 + resistance of light bulb 3 + resistance of light bulb 4

The same current passes throughout a series circuit. Therefore the total current flowing in a series circuit is the same as the individual currents passing through each resistance (or light bulb).

All dry cell batteries and lead–acid batteries used in cars provide a source of electrons. Chemical reactions in these batteries create the electrons. The electrons produced by batteries move in only one direction; this electric current is called direct current. The electricity used in our homes and supplied by utility companies is called alternating current because the electrons move in one direction during the first half of a cycle and in the opposite direction during the second half of the cycle. There are 60 cycles per second.

Electricians use a form of shorthand to represent electrical circuits. Brief shorthand representations of electrical circuits are given in the Reference Table 3 at the beginning of this book. Introduce students to this way of drawing electrical circuits.

In this activity, three dry 1.5 volts "D" cells are used. You can get "D" cell connectors from electronic shops. These connectors provide color–coded connections, eliminating the need for taping the cells in a series. Connecting the negative terminal of a cell to the positive terminal of a second cell, connecting the negative terminal of the second cell to the positive terminal of the third cell, and so on will give a series connection.

A 3 volts light bulb is needed for this activity. The length of the bell wire is not critical.

TEACHER'S SECTION
ELECTRICAL CURRENT IN A SERIES CIRCUIT

GOALS AND OBJECTIVES:

1. To learn that in current electricity, electrons flow.
2. To learn that an electrical circuit describes the path of electron flow.
3. To learn that there is no electron flow in an open circuit.
4. To learn that electrons flow through a closed circuit.
5. To learn that in a series circuit, there is only one path for the electricity to flow through.
6. To learn that in a series circuit, the same current passes throughout the path.
7. To learn that the total current flowing in a series circuit is the same as the individual currents passing through each light bulb.
8. To learn that the total resistance in a series circuit is the sum of the individual resistances in the circuit.

DISCUSSION BEFORE THE LABORATORY ACTIVITY:

It is very important to stress the need for wearing an apron and safety goggles.

Static electricity is electricity at rest. In static electricity, electrical charges are accumulated. Static electricity is produced by rubbing two nonmetallic surfaces so that there is a transfer of electrons from one surface to the other.

Current electricity is the flow of electrons. For the electricity to travel, it must have a path. The complete path through which electricity travels is called a circuit.

There are open circuits and closed circuits. Electrical circuits provide the path for the flow of electrons. Electrons flow from the negative terminal of the battery through the circuit and end at the positive terminal of the battery. Electricity will flow through a closed circuit, because the electrical current can return to its source. Electricity cannot flow through an open circuit, because the electrons coming from the negative terminal cannot reach the positive terminal.

There are series circuits and parallel circuits. In a series circuit, there is only one path for the flow of electricity. If this path has been opened by a loose connection or a loose bulb, electricity will not flow through the circuit. A parallel circuit contains two or more separate paths for the flow of electricity. This will allow the electricity to flow through one path even if the other paths are open.

Resistance is the opposition to the flow of electrons in an electrical circuit. The resistance of a wire depends on the nature and dimensions of the wire.

In a series circuit, the total resistance is the sum of the individual resistances. If the resistance of the connecting wires is not taken into account, then the total resistance is given by the expression:

Total resistance = resistance of light bulb 1 + resistance of light bulb 2 + resistance of light bulb 3 + resistance of light bulb 4

The same current passes throughout a series circuit. Therefore the total current flowing in a series circuit is the same as the individual currents passing through each resistance (or light bulb).

The basic unit of electric current, the ampere is a measure of the number of electrons that move past a point in a conductor in one second. Thus it gives the rate at which electrons move through the conductor.

Electricians use a form of shorthand to represent electrical circuits. Brief shorthand representations of electrical circuits are given in the Reference Table 3 at the beginning of this book. Introduce students to this way of drawing electrical circuits.

In this activity, three dry 1.5 volts "D" cells are used. You can get "D" cell connectors from electronic shops. These connectors provide color-coded connections, eliminating the need for taping the cells in series. Connecting the negative terminal of a cell to the positive terminal of a second cell, connecting the negative terminal of the second cell to the positive terminal of the third cell, and so on will give a series connection.

A 3 volts light bulb is needed for this activity. The length of the bell wire is not critical.

364 Fundamentals of Science Activity Series

TEACHER'S SECTION
PARALLEL CIRCUITS

GOALS AND OBJECTIVES:

1. To learn that in current electricity, electrons flow.
2. To learn that an electrical circuit describes the path of electron flow.
3. To learn that there is no electron flow in an open circuit.
4. To learn that electrons flow through a closed circuit.
5. To learn that two or more paths are available for the electricity to flow in a parallel circuit.
6. To learn that electricity can flow through other branches of a parallel circuit when one branch has been opened by a loose connection or a loose bulb.

DISCUSSION BEFORE THE LABORATORY ACTIVITY:

It is very important to stress the need for wearing an apron and safety goggles.

Static electricity is electricity at rest. In static electricity, electrical charges are accumulated. Static electricity is produced by rubbing two nonmetallic surfaces so that there is a transfer of electrons from one surface to the other.

Current electricity is the flow of electrons. For the electricity to travel, it must have a path. The complete path through which electricity travels is called a circuit.

There are open circuits and closed circuits. Electrical circuits provide the path for the flow of electrons. Electrons flow from the negative terminal of the battery through the circuit and end at the positive terminal of the battery. Electricity will flow through a closed circuit, because the electrical current can return to its source. Electricity cannot flow through an open circuit, because the electrons coming from the negative terminal cannot reach the positive terminal.

There are series circuits and parallel circuits. In a series circuit, there is only one path for the flow of electricity. If this path has been opened by a loose connection or a loose bulb, electricity will not flow through the circuit. A parallel circuit contains two or more separate paths for the flow of electricity. This will allow the electricity to flow through one path even if the other paths are open.

Resistance is the opposition to the flow of electrons in an electrical circuit. The resistance of a wire depends on the nature and dimensions of the wire.

In a parallel circuit, the total resistance is less than the value of any single resistance. For example, if the two resistances are 3 ohms and 6 ohms, the total resistance in the parallel circuit will be less than 3 ohms. The total resistance is given by the expression:

1/Total resistance = 1/resistance of light bulb 1 + 1/resistance of light bulb 2 + 1/resistance of light bulb 3 + 1/resistance of light bulb 4.

The current passing through each branch of a parallel circuit depends on the resistance in that branch. The total current flowing in a parallel circuit is the sum of the individual currents passing through each branch.

Discuss the way to find current and resistance in a series circuit, and compare with a parallel circuit.

The lights, radio, clock, and television in a house are all connected in a parallel circuit. Appliances such as the stove, oven, refrigerator, washer, and dryer also use parallel circuits. When you turn on the switch for a light, only the light on that part of the parallel circuit will be on. The other parts of the parallel circuit will still be open. If there are several light bulbs on a parallel circuit, the remaining light bulbs will still be on, even when one light bulb is burned out. This is because each light bulb has its own independent path to and from the source of electricity.

Electricians use a form of shorthand to represent electrical circuits. Brief shorthand representations of electrical circuits are given in the Reference Table 3 at the beginning of this book. Introduce students to this way of drawing electrical circuits.

In this activity, three dry 1.5 volts "D" cells are used. You can get "D" cell connectors from electronic shops. These connectors provide color–coded connections, eliminating the need for taping the cells in series. Connecting the negative terminal of a cell to the positive terminal of a second cell, connecting the negative terminal of the second cell to the positive terminal of the third cell, and so on will give a series connection.

Four 6 volts lantern light bulbs are needed for this activity. The length of the bell wire is not critical.

TEACHER'S SECTION
ELECTRICAL CURRENT IN A PARALLEL CIRCUIT

GOALS AND OBJECTIVES:

1. To learn that in current electricity, electrons flow.
2. To learn that an electrical circuit describes the path of electron flow.
3. To learn that there is no electron flow in an open circuit.
4. To learn that electrons flow through a closed circuit.
5. To learn that two or more paths are available for the electricity to flow in a parallel circuit.
6. To learn that in a parallel circuit, if one path is open, electricity can flow through other paths that are closed.
7. To learn that the total resistance in a parallel circuit is less than any individual resistance.
8. To learn that the total current in a parallel circuit is the sum of the individual currents flowing through each branch of the circuit.

DISCUSSION BEFORE THE LABORATORY ACTIVITY:

It is very important to stress the need for wearing an apron and safety goggles.

Static electricity is electricity at rest. In static electricity, electrical charges are accumulated. Static electricity is produced by rubbing two nonmetallic surfaces so that there is a transfer of electrons from one surface to the other.

Current electricity is the flow of electrons. For the electricity to travel, it must have a path. The complete path through which electricity travels is called a circuit.

There are open circuits and closed circuits. Electrical circuits provide the path for the flow of electrons. Electrons flow from the negative terminal of the battery through the circuit and end at the positive terminal of the battery. Electricity will flow through a closed circuit, because the electrical current can return to its source. Electricity cannot flow through an open circuit, because the electrons coming from the negative terminal cannot reach the positive terminal.

There are series circuits and parallel circuits. In a series circuit, there is only one path for the flow of electricity. If this path has been opened by a loose connection or a loose bulb, electricity will not flow through the circuit. A parallel circuit contains two or more separate paths for the flow of electricity. This will allow the electricity to flow through one path even if the other paths are open.

Resistance is the opposition to the flow of electrons in an electrical circuit. The resistance of a wire depends on the nature and dimensions of the wire.

In a parallel circuit, the total resistance is less than the value of any single resistance. For example, if the two resistances are 3 ohms and 6 ohms, the total resistance in the parallel circuit will be less than 3 ohms. The total resistance is given by the expression:

1/Total resistance = 1/resistance of light bulb 1 + 1/resistance of light bulb 2 + 1/resistance of light bulb 3 + 1/resistance of light bulb 4.

The current passing through each branch of a parallel circuit depends on the resistance in that branch. The total current flowing in a parallel circuit is the sum of the individual currents passing through each branch.

Discuss the way to find current and resistance in a series circuit, and compare with a parallel circuit.

Electricians use a form of shorthand to represent electrical circuits. Brief shorthand representations of electrical circuits are given in the Reference Table 3 at the beginning of this book. Introduce students to this way of drawing electrical circuits.

In this activity, three dry 1.5 volts "D" cells are used. You can get "D" cell connectors from electronic shops. These connectors provide color-coded connections, eliminating the need for taping the cells in series. Connecting the negative terminal of a cell to the positive terminal of a second cell, connecting the negative terminal of the second cell to the positive terminal of the third cell, and so on will give a series connection.

Four 6 volts lantern light bulbs are needed for this activity. The length of the bell wire is not critical.

TEACHER'S SECTION
MAKING A LIGHT BULB

GOALS AND OBJECTIVES:

1. To learn the principle of an incandescent light bulb.
2. To learn that nichrome wire offers much resistance to the flow of electricity.
3. To learn that the resistance of a wire depends on the nature of the wire and the dimensions of the wire.
4. To learn that electrical energy can be changed into heat and light.
5. To learn that the brightness of a light bulb depends on the dissipation of electrical energy as heat and light.
6. To learn that the unit for measuring the resistance of a wire is called an ohm.

DISCUSSION BEFORE THE LABORATORY ACTIVITY:

It is very important to stress the need for wearing an apron and safety goggles.

Discuss the meaning of electrical circuits, open circuits, and closed circuits. Electrical circuits provide the path for the flow of electrons. Electricity will flow only if it can return to its source.

Metals such as copper and aluminum are good conductors of electricity. Copper wires are used for the electrical wiring in homes. These copper wires are insulated with plastic, because plastic does not conduct electricity. This allows you to touch the insulated wire even when the electricity is flowing through the wire. Insulated copper wires of different thicknesses are used, depending on the amount of current expected to flow through them. Thicker wires can carry more electricity without getting hot.

The ease with which electrons flow through a metal depends on the nature of the metal. The wires used in electrical wiring conduct electricity with ease. The wires used in making light bulbs, electrical hot plates, toaster ovens, and space heating offer a great deal of resistance for the flow of electricity, and they dissipate electrical energy in the form of heat.

Resistance is a property of conductors. The current produced by a given potential difference (voltage) is determined by the resistance in the circuit. The resistance of a wire is determined by the nature, the dimensions, and temperature of the wire.

The practical unit of resistance is called an ohm. One ohm is the amount of resistance that will produce a current of 1 ampere when a potential difference of 1 volt is applied to the circuit. The resistance values of 100 feet of standard annealed copper at 20° Celsius are 0.1588 ohm for American Wire gauge No. 12, 0.2525 ohm for AWG No. 14, and 0.6385 ohm for AWG No. 18. These copper wires are often used for wiring in homes. The resistance values of 100 feet of nichrome wire are 10.29, 16.48, 42.19, and 675 ohms for American Wire Gauge Numbers 12, 14, 18, and 30 respectively. Thus nichrome offers much more resistance for the flow of electricity. Nichrome dissipates a lot of electrical energy in the form of heat and light.

The resistance of a 1 centimeter cube of a substance to the flow of electricity is known as the specific resistance or resistivity. The current flow is perpendicular to two parallel faces of the cube. This resistivity is characteristic of different substances. It is defined by the expression:

Resistance = resistivity x length/cross sectional area

where the resistance of a uniform conductor is measured in a unit called ohms, the length of the conductor is measured in centimeters; the cross-sectional area of the conductor is measured in square centimeters; and the resistivity of the conductor is measured in a unit called ohm centimeters.

At 20° Celsius, the resistivity values of annealed copper and nichrome are 0.0000017241 and 0.000100 ohm centimeters respectively.

Thomas Edison invented the electric light in the late 1870s. Electric arc lamps were well known before the invention of incandescent lamps. The biggest problem in developing the incandescent lamp was to get a material that would not melt or decompose under intense heat. Edison experimented with a variety of lights, using platinum wire and carbonized cotton thread.

A modern incandescent light uses a coil of tungsten filament in an inert gas (argon) atmosphere. The argon reduces the evaporation of the filament. This filament gets extremely hot and glows without melting. The temperature of this filament is about 3000°Celsius when it is glowing "white hot." Incandescent light bulbs with a wide range of power are commercially available.

Discuss other sources of lights, such as carbon arc lamps, fluorescent light bulbs, sodium vapor lamps used in highway lighting, neon lights, ultraviolet lights, and laser lights.

Fluorescent lamps give off radiant energy. These tube shaped–lamps are coated with special fluorescent material and contain mercury vapor. The flow of electricity causes mercury to produce ultraviolet radiation. This radiation hits the coated surface, which, in turn, produces the glow. Fluorescent lamps use less electricity and last longer. They do not waste energy as heat. However their purchase price is more than the incandescent light bulbs.

In this activity, three dry 1.5 volts "D" cells are used. You can get "D" cell connectors from electronic shops. These connectors provide color–coded connections, eliminating the need for taping the cells in a series. Connecting the negative terminal of a cell to the positive terminal of a second cell, connecting the negative terminal of the second cell to the positive terminal of the third cell, and so on will give a series connection.

In this activity, it is better to use nichrome wire with the smallest thickness.

TEACHER'S SECTION
MAGNETIC AND NONMAGNETIC OBJECTS

GOALS AND OBJECTIVES:

1. To learn that some materials are attracted by a magnet.
2. To learn that some other materials are not attracted by a magnet.
3. To learn to distinguish between magnetic and nonmagnetic objects.
4. To observe that objects made of iron are attracted by a magnet.
5. To observe that nonmetallic objects are also nonmagnetic.

DISCUSSION BEFORE THE LABORATORY ACTIVITY:

It is very important to stress the need for wearing an apron and safety goggles.

It was discovered a long time ago that lodestone, a rock found naturally on earth, attracts small pieces of iron. Lodestone is now known to chemists as a compound of iron oxide. Found near the ancient city of Magnesia, Chinese sailors discovered that rubbing a lodestone with an iron needle made the needle attract iron pieces. They also discovered that such a needle, when suspended freely, always points north.

A compass contains a magnetized needle that can turn freely. The needle always points north no matter how the compass is turned.

A magnet is a substance that attracts iron. Iron, steel, cobalt, and nickel are all magnetic. Metals such as copper, zinc, and gold are not magnetic. Nonmetals such as paper and glass are not magnetic. The liquid made from the oxygen gas we breathe is very special. It is magnetic.

Permanent magnets of differing size, shape, and strength are made of steel. The most common ones are bar magnets, horseshoe magnets, and circular magnets. Very powerful ceramic magnets, alnico (an alloy of iron, aluminum, nickel and cobalt) magnets and Neo (an alloy of neodymium, iron, and boron) magnets are also available commercially. Ceramic magnets are stronger than alnico magnets. Neo magnets are much stronger than ceramic magnets.

TEACHER'S SECTION
MAGNETIC ATTRACTION THROUGH OTHER MATERIALS

GOALS AND OBJECTIVES:

1. To learn that some materials are attracted by a magnet.
2. To learn that some other materials are not attracted by a magnet.
3. To learn to distinguish between magnetic and nonmagnetic objects.
4. To observe that objects made of iron are attracted by a magnet.
5. To observe that nonmetallic objects are also nonmagnetic.
6. To learn that a magnetic force of attraction can pass through a nonmagnetic object.
7. To learn that a magnetic force from a magnet gets concentrated in a magnetic object and this object will not let the magnetic force pass through it.

DISCUSSION BEFORE THE LABORATORY ACTIVITY:

It is very important to stress the need for wearing an apron and safety goggles.

A magnet is a substance that attracts iron. Iron, steel, cobalt, and nickel are all magnetic. Metals such as copper, zinc, and gold are not magnetic. Nonmetals such as paper and glass are not magnetic. The liquid made from the oxygen gas we breathe is very special. It is magnetic.

A compass contains a magnetized needle that can turn freely. The needle always points north no matter how the compass is turned.

The force with which a magnet attracts other magnetic objects depends on the distance between the two. The closer they are the stronger the attraction. As the distance between the magnet and the magnetic object increases, the attraction decreases.

Permanent magnets of differing size, shape, and strength are made of steel. The most common ones are bar magnets, horseshoe magnets, and circular magnets. Very powerful ceramic magnets, alnico (an alloy of iron, aluminum, nickel and cobalt) magnets and Neo (an alloy of neodymium, iron, and boron) magnet are also available commercially. Ceramic magnets are stronger than alnico magnets. Neo magnets are much stronger than ceramic magnets.

The force of attraction exerted by a magnet can pass through nonmagnetic materials such as paper, clay, wood, air, and glass.

Materials such as iron and steel are highly magnetic. They concentrate the magnetic force through them. They will not let a magnetic force pass through them. You can move a metallic paper clip kept on a paper by moving a magnet under the paper. However you cannot move a metallic paper clip kept in a tin can by moving a magnet under the tin can.

TEACHER'S SECTION
NORTH POLE AND SOUTH POLE OF A MAGNET

GOALS AND OBJECTIVES:

1. To learn that a magnet has two poles.
2. To learn that the poles of a magnet are called the north pole and the south pole.
3. To observe that a magnetic object, when brought near the north pole of a magnet, is attracted by the pole.
4. To observe that a magnetic object, when brought near the south pole of a magnet, is attracted by the pole.
5. To observe that when the north pole of a magnet, is brought near the north pole of another suspended magnet, they repel each other, and the north pole of the second suspended magnet is pushed away from the north pole of the first magnet.
6. To observe that the south pole of a magnet, when brought near the north pole of another suspended magnet, the two poles attract each other.
7. To learn that like poles of two magnets repel each other.
8. To learn that unlike poles of two magnets attract each other.
9. To learn that a magnetic object and a magnet respond differently to the two poles of a suspended magnet.

DISCUSSION BEFORE THE LABORATORY ACTIVITY:

It is very important to stress the need for wearing an apron and safety goggles.

A magnet is a substance that attracts iron. Iron, steel, cobalt, and nickel are all magnetic. Metals such as copper, zinc, and gold are not magnetic. Nonmetals such as paper and glass are not magnetic. The liquid made from the oxygen gas we breathe is very special. It is magnetic.

The force with which a magnet attracts another magnetic object depends on the distance between the two. The closer they are, the stronger the attraction. As the distance between the magnet and the magnetic object increases, the attraction decreases.

It was discovered a long time ago that lodestone, a rock found naturally on earth, attracts small pieces of iron. Lodestone is now known to chemists as a compound of iron oxide. Found near the ancient city of Magnesia, Chinese sailors discovered that rubbing a lodestone with an iron needle made the needle attract iron pieces. They also discovered that such a needle, when suspended freely, always points north.

A compass contains a magnetized needle that can turn freely. The needle always points north no matter how the compass is turned. The two ends of any freely suspended magnet will point towards the north and south. The end of a suspended magnet that points towards the north is called the north pole of the magnet. The end of the suspended magnet that points towards the south is called the south pole of the magnet. All magnets have two poles, north pole and south pole. When a magnet is broken into several pieces, each piece will act like a magnet, and each piece will have both north and south poles. Each piece of the magnet will attract other nearby magnetic objects through its invisible field of force. The invisible field of force surrounding every magnet is strongest at the poles of the magnet.

The earth itself acts as a magnet. The geographic north and south poles of the earth are on the axis of rotation. The magnetic north pole and south pole do not coincide with the geographic north pole and south pole. The magnetization axis makes an angle of about 15 degrees with the earth's axis. The entire earth is assumed to be a magnet with two poles. It is assumed that the north pole of the earth's magnet is distributed over the entire southern hemisphere, and the south pole of the earth's magnet is distributed over the entire northern hemisphere. Thus the north pole of the earth's magnet is directed toward the general geographic south pole and the south pole of the earth's magnet is directed towards the general geographic north pole.

When a magnet is suspended freely, its north pole is attracted by the south pole of the earth's magnet, and hence the north pole of the magnet points in the general direction of geographic north. The south pole of the suspended magnet is attracted by the north pole of the earth's magnet, and hence the south pole of the magnet points in the general direction of geographic south.

TEACHER'S SECTION
MAGIC WITH MAGNETS

GOALS AND OBJECTIVES:

1. To learn that like poles of two magnets repel each other.

2. To learn that unlike poles of two magnets attract each other.

3. To learn that a magnetic object cannot be suspended in space on top of another magnetic object.

4. To observe that a magnet can be suspended in space on top of another magnet.

5. To learn that if the magnetic force of repulsion between two like poles is greater than the downward gravitational force, a magnet can be suspended in space over the top of another magnet.

DISCUSSION BEFORE THE LABORATORY ACTIVITY:

It is very important to stress the need for wearing an apron and safety goggles.

It was discovered a long time ago that lodestone, a rock found naturally on earth, attracts small pieces of iron. Lodestone is now known to chemists as a compound of iron oxide. Found near the ancient city of Magnesia, Chinese sailors discovered that rubbing a lodestone with an iron needle made the needle attract iron pieces. They also discovered that such a needle, when suspended freely, always points north.

A compass contains a magnetized needle that can turn freely. The needle always points north no matter how the compass is turned. The two ends of any freely suspended magnet will point towards the north and south. The end of a suspended magnet that points towards the north is called the north pole of the magnet. The end of the suspended magnet that points towards the south is called the south pole of the magnet. All magnets have two poles: north pole and south pole. When a magnet is broken into several pieces, each piece will act like a magnet, and each piece will have both north and south poles. Each piece of the magnet will attract other nearby magnetic objects through its invisible field of force. The invisible field of force surrounding every magnet is strongest at the poles of the magnet.

The earth itself acts as a magnet. The geographic north and south poles of the earth are on the axis of rotation. The magnetic north pole and south pole do not coincide with the geographic north pole and south pole. The magnetization axis makes an angle of about 15 degrees with the earth's axis. The entire earth is assumed to be a magnet with two poles. It is assumed that the north pole of the earth's magnet is distributed over the entire southern hemisphere, and the south pole of the earth's magnet is distributed over the entire northern hemisphere. Thus the north pole of the earth's magnet is directed toward the general geographic south pole and the south pole of the earth's magnet is directed towards the general geographic north pole.

When a magnet is suspended freely, its north pole is attracted by the south pole of the earth's magnet, and hence the north pole of the magnet points in the general direction of geographic north. The south pole of the suspended magnet is attracted by the north pole of the earth's magnet, and hence the south pole of the magnet points in the general direction of geographic south.

The force with which a magnet attracts other magnetic objects depends on the distance between the two. The closer they are, the stronger the attraction. As the distance between the magnet and the magnetic object increases, the attraction decreases.

There is attraction between two unlike poles of two magnets. There is repulsion between two like poles of two magnets. Thus two north poles or two south poles of two magnets repel each other. A north pole of one magnet attracts the south pole of another magnet. The force with which a magnetic pole attracts another magnetic pole or a magnetic pole repels another magnetic pole depends on the distance between the two poles. The closer the two unlike poles, the stronger the attraction. The closer the two like poles, the stronger the repulsion.

When an object is thrown upward, it does not remain suspended in space, because of the downward gravitational force of attraction between the object and the earth. The bigger the mass of the object, the greater the gravitational force of attraction.

Whether one can pack two or more magnets one on top of the other depends on (1) how they are packed and (2) on the masses of the magnets. If the magnets are packed in such a way that their like poles face each other, it may be possible to suspend them in space, one on top of the other, provided the downward gravitational force of attraction is less than the repulsion between two like poles.

TEACHER'S SECTION
MAGNETIZING OTHER OBJECTS

GOALS AND OBJECTIVES:

1. To learn that only magnetic materials can be turned into magnets.
2. To learn that in order to make a magnet, the magnetic field from a magnet must pass through the magnetic material.
3. To learn that the flow of magnetic field through a material causes a change in the direction of magnetic fields of atoms inside the object.
4. To learn that an object can be magnetized by stroking it against one of the poles of a magnet in one direction.
5. To learn that a temporary magnet can be made, using soft iron.
6. To learn that a permanent magnet can be made, using steel.
7. To learn that an electromagnet retains its magnetic properties only when the electricity is turned on.
8. To learn that electromagnets are temporary magnets.

DISCUSSION BEFORE THE LABORATORY ACTIVITY:

It is very important to stress the need for wearing an apron and safety goggles.

It was discovered a long time ago that lodestone, a rock found naturally on earth, attracts small pieces of iron. Lodestone is now known to chemists as a compound of iron oxide. Found near the ancient city of Magnesia, Chinese sailors discovered that rubbing a lodestone with an iron needle made the needle attract iron pieces. They also discovered that such a needle, when suspended freely, always points north.

A compass contains a magnetized needle that can turn freely. The needle always points north no matter how the compass is turned. The two ends of any freely suspended magnet will point towards the north and south. The end of a suspended magnet that points towards the north is called the north pole of the magnet. The end of the suspended magnet that points towards the south is called the south pole of the magnet. All magnets have two poles, north pole and south pole. When a magnet is broken into several pieces, each piece will act like a magnet, and each piece will have both north and south poles. Each piece of the magnet will attract other nearby magnetic objects through its invisible field of force. The invisible field of force surrounding every magnet is strongest at the poles of the magnet.

A magnet is a substance that attracts iron. Iron, steel, cobalt, and nickel are all magnetic. Metals such as copper, zinc, and gold are not magnetic. Nonmetals such as paper and glass are not magnetic.

The earth itself acts as a magnet. The geographic north and south poles of the earth are on the axis of rotation. The magnetic north pole and south pole do not coincide with the geographic north pole and south pole. The magnetization axis makes an angle of about 15 degrees with the earth's axis. The entire earth is assumed to be a magnet with two poles. It is assumed that the north pole of the earth's magnet is distributed over the entire southern hemisphere, and the south pole of the earth's magnet is distributed over the entire northern hemisphere. Thus the north pole of the earth's magnet is directed toward the general geographic south pole and the south pole of the earth's magnet is directed towards the general geographic north pole.

It is possible to use earth's magnetism to produce other magnets.

Temporary magnets are made from soft iron. Permanent magnets are made from steel.

Permanent magnets of differing size, shape, and strength are made of steel. The most common ones are bar magnets, horseshoe magnets, and circular magnets. Very powerful ceramic magnets, alnico (an alloy of iron, aluminum, nickel and cobalt) magnets, and Neo (an alloy of neodymium, iron, and boron) magnets also are available commercially. Ceramic magnets are stronger than alnico magnets. Neo magnets are much stronger than ceramic magnets.

Only a magnetic material can be changed into a magnet. When a magnetic object is stroked against one of the poles of a magnet in one direction, the magnetic field in the atoms of the object gets aligned in one direction and this alignment turns the object into a magnet.

An electromagnet is a temporary magnet, whose core is made of soft iron. It will behave like a magnet only when electricity is passing through the coil surrounding the iron core.

TEACHER'S SECTION
MAKING A COMPASS

GOALS AND OBJECTIVES:

1. To learn that only magnetic materials can be turned into magnets.
2. To learn that in order to make a magnet, the magnetic field from a magnet must pass through the magnetic material.
3. To learn that the flow of magnetic field through a material causes a change in the direction of the magnetic fields of atoms inside the object.
4. To learn that an object can be magnetized by stroking it against one of the poles of a magnet in one direction.
5. To learn that a permanent magnet can be made from steel.
6. To learn that a compass is made from a strong magnet.
7. To learn that the needle of a compass points in the generally north and south directions.

DISCUSSION BEFORE THE LABORATORY ACTIVITY:

It is very important to stress the need for wearing an apron and safety goggles.

It was discovered a long time ago that lodestone, a rock found naturally on earth, attracts small pieces of iron. Lodestone is now known to chemists as a compound of iron oxide. Found near the ancient city of Magnesia, Chinese sailors discovered that rubbing a lodestone with an iron needle made the needle attract iron pieces. They also discovered that such a needle, when suspended freely, always points north.

A compass contains a magnetized needle that can turn freely. The needle always points north no matter how the compass is turned. The two ends of any freely suspended magnet will point towards the north and south. The end of a suspended magnet that points towards the north is called the north pole of the magnet. The end of the suspended magnet that points towards the south is called the south pole of the magnet. All magnets have two poles: north pole and south pole. When a magnet is broken into several pieces, each piece will act like a magnet, and each piece will have both north and south poles. Each piece of the magnet will attract other nearby magnetic objects through its invisible field of force. The invisible field of force surrounding every magnet is strongest at the poles of the magnet.

A magnet is a substance that attracts iron. Iron, steel, cobalt, and nickel are all magnetic. Metals such as copper, zinc, and gold are not magnetic. Nonmetals such as paper and glass are not magnetic.

The earth itself acts as a magnet. The geographic north and south poles of the earth are on the axis of rotation. The magnetic north pole and south pole do not coincide with the geographic north pole and south pole. The magnetization axis makes an angle of about 15 degrees with the earth's axis. The entire earth is assumed to be a magnet with two poles. It is assumed that the north pole of the earth's magnet is distributed over the entire southern hemisphere, and the south pole of the earth's magnet is distributed over the entire northern hemisphere. Thus the north pole of the earth's magnet is directed toward the general geographic South Pole and the south pole of the earth's magnet is directed towards the general geographic north pole.

It is possible to use earth's magnetism to produce other magnets.

Only a magnetic material can be changed into a magnet. When a magnetic object is stroked against one of the poles of a magnet in one direction, the magnetic field in the atoms of the object gets aligned in one direction and this alignment turns the object into a magnet.

TEACHER'S SECTION
STRENGTH OF A MAGNET

GOALS AND OBJECTIVES:

1. To learn that a magnet has two poles.//
2. To learn that the poles of a magnet are called north pole and south pole.
3. To observe that a magnetic object, when brought near the north pole of a magnet, is attracted by the pole.
4. To observe that a magnetic object, when brought near the south pole of a magnet, is attracted by the pole.
5. To learn that the strength of the north pole is the same as the strength of the south pole of a magnet.
6. To learn that the poles of a magnet have maximum strength.
7. To learn that the center of a magnet has minimum strength.
8. To learn that different magnets can have different strengths.

DISCUSSION BEFORE THE LABORATORY ACTIVITY:

It is very important to stress the need for wearing an apron and safety goggles.

A magnet is a substance that attracts iron. Iron, steel, cobalt, and nickel are all magnetic. These materials are pulled by a force called magnetism. This force can be used to do useful work.

The force with which a magnet attracts or pulls other magnetic objects depends on the distance between the two. The closer they are, the stronger the attraction or pull. As the distance between the magnet and the magnetic object increases, the attraction or pull decreases. The field of force surrounding a magnet is invisible.

It was discovered a long time ago that lodestone, a rock found naturally on earth, attracts small pieces of iron. Lodestone is now known to chemists as a compound of iron oxide. Found near the ancient city of Magnesia, Chinese sailors discovered that rubbing a lodestone with an iron needle made the needle attract iron pieces. They also discovered that such a needle, when suspended freely, always points north.

A compass contains a magnetized needle that can turn freely. The needle always points north no matter how the compass is turned. The two ends of any freely suspended magnet will point toward the north and south. The end of a suspended magnet that points towards the north is called the north pole of the magnet. The end of the suspended magnet that points towards the south is called the south pole of the magnet. All magnets have two poles: north pole and south pole.

The earth itself acts as a magnet. The geographic north and south poles of the earth are on the axis of rotation. The magnetic north pole and south pole do not coincide with the geographic north pole and south pole. The magnetization axis makes an angle of about 15 degrees with the earth's axis. The entire earth is assumed to be a magnet with two poles. It is assumed that the north pole of the earth's magnet is distributed over the entire southern hemisphere, and the south pole of the earth's magnet is distributed over the entire northern hemisphere. Thus the north pole of the earth's magnet is directed toward the general geographic South Pole and the south pole of the earth's magnet is directed towards the general geographic north pole.

Permanent magnets of differing size, shape, and strength are made of steel. The most common ones are bar magnets, horseshoe magnets, and circular magnets. Very powerful ceramic magnets, alnico (an alloy of iron, aluminum, nickel and cobalt) magnets, and Neo (an alloy of neodymium, iron, and boron) magnets are also available commercially. Ceramic magnets are stronger than alnico magnets. Neo magnets are much stronger than ceramic magnets.

A magnet is strongest at the two poles. The north pole and south pole of a magnet have the same strength. The center of a magnet has the minimum strength. The strength of a magnet can be found by determining the amount of force exerted by a magnet against gravitational force. If a magnet can support numerous iron pieces or metallic paper clips and keep them suspended against gravity, then that magnet has a lot of strength.

TEACHER'S SECTION
MAGNETIC FIELD AROUND A MAGNET

GOALS AND OBJECTIVES:

1. To learn that a magnet has two poles.
2. To learn that the poles of a magnet are called north pole and south pole.
3. To observe that iron filings crowd around the poles of a magnet.
4. To learn that the magnetic field around a magnet is strongest near the poles and weakest near the middle of the magnet.
5. To learn that the magnetic field around a magnet becomes weaker with increasing distance from the poles.
6. To learn that the magnetic field around a magnet originates from the north pole and ends in the south pole.

DISCUSSION BEFORE THE LABORATORY ACTIVITY:

It is very important to stress the need for wearing an apron and safety goggles.

A magnet is a substance that attracts iron. Iron, steel, cobalt, and nickel are all magnetic. These materials are pulled by a force called magnetism. This force can be used to do useful work.

The force with which a magnet attracts or pulls other magnetic objects depends on the distance between the two. The closer they are, the stronger the attraction or pull. As the distance between the magnet and the magnetic object increases, the attraction or pull decreases. The field of force surrounding a magnet is invisible.

It was discovered a long time ago that lodestone, a rock found naturally on earth, attracts small pieces of iron. Lodestone is now known to chemists as a compound of iron oxide. Found near the ancient city of Magnesia, Chinese sailors discovered that rubbing a lodestone with an iron needle made the needle attract iron pieces. They also discovered that such a needle, when suspended freely, always points north.

The earth itself acts as a magnet. The geographic north and south poles of the earth are on the axis of rotation. The magnetic north pole and south pole do not coincide with the geographic north pole and south pole. The magnetization axis makes an angle of about 15 degrees with the earth's axis. The entire earth is assumed to be a magnet with two poles. It is assumed that the north pole of the earth's magnet is distributed over the entire southern hemisphere, and the south pole of the earth's magnet is distributed over the entire northern hemisphere. Thus the north pole of the earth's magnet is directed toward the general geographic south pole and the south pole of the earth's magnet is directed towards the general geographic north pole.

When a magnet is suspended freely, its north pole is attracted by the south pole of the earth's magnet, so the north pole of the magnet points in the general direction of geographic north. The south pole of the suspended magnet is attracted by the north pole of the earth's magnet, and hence the south pole of the magnet points in the general direction of geographic south.

A magnet has a magnetic field or magnetic lines of force around it. The magnetic field around the magnet originates from the north pole and ends in the south pole. Inside the magnet, the magnetic field is from the south pole to the north pole.

When iron filings are sprinkled around a magnet, they get magnetized by induction and they orient themselves in the direction of the magnetic field. Because the magnetic field is strongest near the poles, the iron filings get crowded near the poles. The field near the north pole and the field near the south pole are of equal strength.

TEACHER'S SECTION
MAGNETIC FIELD BETWEEN TWO UNLIKE POLES

GOALS AND OBJECTIVES:

1. To learn that a magnet has two poles.
2. To learn that the poles of a magnet are called north pole and south pole.
3. To observe that iron filings crowd around the poles of a magnet.
4. To learn that the magnetic field around a magnet is strongest near the poles and weakest near the middle of the magnet.
5. To learn that the magnetic field around a magnet becomes weaker with increasing distance from the poles.
6. To learn that the magnetic field around a magnet originates from the north pole and ends in the south pole.
7. To learn that two unlike poles of two magnets attract each other.
8. To learn that there are maximum magnetic lines of force or magnetic field between two unlike poles of two magnets.
9. To learn that iron filings get crowded between two unlike poles of two magnets.

DISCUSSION BEFORE THE LABORATORY ACTIVITY:

It is very important to stress the need for wearing an apron and safety goggles.

A magnet is a substance that attracts iron. Iron, steel, cobalt, and nickel are all magnetic. These materials are pulled by a force called magnetism. This force can be used to do useful work.

The force with which a magnet attracts or pulls other magnetic objects depends on the distance between the two. The closer they are, the stronger the attraction or pull. As the distance between the magnet and the magnetic object increases, the attraction or pull decreases. The field of force surrounding a magnet is invisible.

When a magnet is suspended freely, its north pole is attracted by the south pole of the earth's magnet, so the north pole of the magnet points in the general direction of geographic north. The south pole of the suspended magnet is attracted by the north pole of the earth's magnet, and hence the south pole of the magnet points in the general direction of geographic south.

A magnet has a magnetic field or magnetic lines of force around it. The magnetic field around the magnet originates from the north pole and ends in the south pole. Inside the magnet, the magnetic field is from the south pole to the north pole.

When iron filings are sprinkled around a magnet, they get magnetized by induction and they orient themselves in the direction of the magnetic field. Because the magnetic field is strongest near the poles, the iron filings get crowded near the poles. The field near the north pole and the field near the south pole are of equal strength.

When two bar magnets are placed in a line with some distance between their poles, the magnetic lines of force between them depends on whether the two poles are facing each other. If the north pole of one magnet is facing the south pole of the second magnet, then these two unlike poles attract each other. There will be maximum lines of force or magnetic field between these two unlike poles. This can be demonstrated by sprinkling iron filings. Iron filings will become concentrated between the two unlike poles.

TEACHER'S SECTION
MAGNETIC FIELD BETWEEN TWO LIKE POLES

GOALS AND OBJECTIVES:

1. To learn that the poles of a magnet are called north pole and south pole.

2. To observe that iron filings crowd around the poles of a magnet.

3. To learn that the magnetic field around a magnet is strongest near the poles and weakest near the middle of the magnet.

4. To learn that the magnetic field around a magnet becomes weaker with increasing distance from the poles.

5. To learn that the magnetic field around a magnet originates from the north pole and ends in the south pole.

6. To learn that two like poles of two magnets repel each other.

7. To learn that there are minimum magnetic lines of force or magnetic field between two like poles of two magnets.

8. To learn that iron filings will be less dense between two like poles of two magnets.

DISCUSSION BEFORE THE LABORATORY ACTIVITY:

It is very important to stress the need for wearing an apron and safety goggles.

A magnet is a substance that attracts iron.

The force with which a magnet attracts iron filings depends on the distance between the magnet and the iron filings. The closer they are the stronger the attraction. As the distance between the magnet and the iron filings increases, the attraction decreases.

When a magnet is suspended freely, its north pole is attracted by the south pole of the earth's magnet, so the north pole of the magnet points in the general direction of geographic north. The south pole of the suspended magnet is attracted by the north pole of the earth's magnet, and hence the south pole of the magnet points in the general direction of geographic south.

A magnet has a magnetic field or magnetic lines of force around it. The magnetic field around the magnet originates from the north pole and ends in the south pole. Inside the magnet, the magnetic field is from the south pole to the north pole.

When iron filings are sprinkled around a magnet, they get magnetized by induction and they orient themselves in the direction of the magnetic field. Because the magnetic field is strongest near the poles, the iron filings get crowded near the poles. The field near the north pole and the field near the south pole are of equal strength.

When two bar magnets are placed in a line with some distance between their poles, the magnetic lines of force between them depends on whether the two poles are facing each other. If the north pole of one magnet is facing the north pole of the second magnet, then these two like poles repel each other. There will be minimum lines of force or magnetic field between these two like poles. This can be demonstrated by sprinkling iron filings. Iron filings do not concentrate between the two like poles.

TEACHER'S SECTION
MAPPING A MAGNETIC FIELD USING A COMPASS

GOALS AND OBJECTIVES:

1. To learn that the magnetic lines of force converge near the poles of a magnet.

2. To learn that the magnetic lines of force spread out near the center of a magnet.

3. To learn that the direction of the magnetic lines of force are parallel to the magnet near the center of the magnet.

4. To learn that the direction of the magnetic lines of force is away from the magnet at the north pole.

5. To learn that the direction of the magnetic lines of force is toward the magnet at the south pole.

6. To learn that the closeness of the magnetic lines of force indicates the strength of the magnetic field.

7. To learn that the magnetic field is weakest near the center of the magnet.

8. To learn that the magnetic field is strongest near the poles of the magnet.

DISCUSSION BEFORE THE LABORATORY ACTIVITY:

It is very important to stress the need for wearing an apron and safety goggles.

A magnet is a substance that attracts iron. Iron, steel, cobalt, and nickel are all magnetic. Metals such as copper, zinc, and gold are not magnetic. Nonmetals such as paper and glass are not magnetic.

The force with which a magnet attracts other magnetic objects depends on the distance between the two. The closer they are, the stronger the attraction. As the distance between the magnet and the magnetic object increases, the attraction decreases.

It was discovered a long time ago that lodestone, a rock found naturally on earth, attracts small pieces of iron. Lodestone is now known to chemists as a compound of iron oxide. Found near the ancient city of Magnesia,

Chinese sailors discovered that rubbing a lodestone with an iron needle made the needle attract iron pieces. They also discovered that such a needle, when suspended freely, always points north.

The earth itself acts as a magnet. The geographic north and south poles of the earth are on the axis of rotation. The magnetic north pole and south pole do not coincide with the geographic north pole and south pole. The magnetization axis makes an angle of about 15 degrees with the earth's axis. The entire earth is assumed to be a magnet with two poles. It is assumed that the north pole of the earth's magnet is distributed over the entire southern hemisphere, and the south pole of the earth's magnet is distributed over the entire northern hemisphere. Thus the north pole of the earth's magnet is directed toward the general geographic south pole and the south pole of the earth's magnet is directed towards the general geographic north pole.

When a magnet is suspended freely, its north pole is attracted by the south pole of the earth's magnet, and hence the north pole of the magnet points in the general direction of geographic north. The south pole of the suspended magnet is attracted by the north pole of the earth's magnet, and hence the south pole of the magnet points in the general direction of geographic south.

A magnet has a magnetic field or magnetic lines of force around it. The magnetic field around the magnet originates from the north pole and ends in the south pole. Inside the magnet, the magnetic field is from the south pole to the north pole.

When iron filings are sprinkled around a magnet, these filings get magnetized by induction and they orient themselves in the direction of the magnetic field. Because the magnetic field is strongest near the poles, the iron filings get crowded near the poles. The field near the north pole and the field near the south pole are of equal strength.

The magnetic field or magnetic lines of force can also be observed by using a compass. The needle of the compass aligns itself in the direction of the filed. Thus the south pole of a compass will point towards the north pole of a magnet. The north pole of the compass will be in the direction of the magnetic field. The magnetic lines of force are strongest near the poles and weakest near the center of a magnet. The direction of the magnetic lines of force is parallel to the magnet near the center. The direction of the magnetic lines of force is away from the north pole and toward the south pole.

Discuss the advantages and disadvantages of using a compass instead of iron filings to observe the magnetic field around a magnet.

TEACHER'S SECTION
MAGNETISM FROM ELECTRICITY

GOALS AND OBJECTIVES:

1. To learn that magnetism can be produced from electricity.

2. To learn that a flow of electrons along a wire produces a magnetic field around the wire.

3. To learn that in order to produce magnetic effects without a magnet, one needs electricity.

4. To learn that when electricity is flowing through a wire, the direction of the magnetic field on top of the wire is different from the direction of the magnetic field at the bottom of the wire.

5. To learn that the direction of the magnetic field around a current carrying wire can be determined by using a compass.

6. To learn that iron filings cling to a current carrying wire because of the magnetic field around the wire.

7. To learn that an electromagnet retains its magnetic properties only when the electricity is turned on.

8. To learn that electromagnets are temporary magnets.

DISCUSSION BEFORE THE LABORATORY ACTIVITY:

It is very important to stress the need for wearing an apron and safety goggles.

It was discovered a long time ago that lodestone, a rock found naturally on earth, attracts small pieces of iron. Lodestone is now known to chemists as a compound of iron oxide. Found near the ancient city of Magnesia, Chinese sailors discovered that rubbing a lodestone with an iron needle made the needle attract iron pieces. They also discovered that such a needle, when suspended freely, always points north.

A compass contains a magnetized needle that can turn freely. The needle always points north, no matter how the compass is turned.

A magnet is a substance that attracts iron. Iron, steel, cobalt, and nickel are all magnetic. Metals such as copper, zinc, and gold are not magnetic. Nonmetals such as paper and glass are not magnetic.

The earth itself acts as a magnet. The geographic north and south poles of the earth are on the axis of rotation. The magnetic north pole and south pole do not coincide with the geographic north pole and south pole. The magnetization axis makes an angle of about 15 degrees with the earth's axis. The entire earth is assumed to be a magnet with two poles. It is assumed that the north pole of the earth's magnet is distributed over the entire southern hemisphere, and the south pole of the earth's magnet is distributed over the entire northern hemisphere. Thus the north pole of the earth's magnet is directed toward the general geographic south pole and the south pole of the earth's magnet is directed towards the general geographic north pole.

It is possible to use earth's magnetism to produce other magnets.

Temporary magnets are made using soft iron. Permanent magnets of differing size, shape and strength are made of steel. The most common ones are bar magnets, horseshoe magnets, and circular magnets. Very powerful ceramic magnets, alnico (an alloy of iron, aluminum, nickel and cobalt) magnets, and Neo (an alloy of neodymium, iron, and boron) magnets are also available commercially. Ceramic magnets are stronger than alnico magnets. Neo magnets are much stronger than ceramic magnets.

An electromagnet is a temporary magnet with a core made of soft iron. It will behave like a magnet only when electricity is passing through the coil surrounding the iron core. Very powerful electromagnets are made using electricity. The great advantage of the electromagnet is that it will act as a magnet only when the electricity is turned on. It loses its magnetic properties when the electricity is turned off.

Electricity and magnetism are closely related to each other, but they are not the same. Both electricity and magnetism involve the movement of electrons. Electricity can be used to produce magnetism, and magnetism can be used to produce electricity.

In this activity, two 1.5 volts dry "D" cells are used. You can get "D" cell connectors from an electronic shop. This connector provides color-coded connections, eliminating the need for taping the cells in a series. Connecting the negative terminal of a cell to the positive terminal of a second cell will give a series connection.

TEACHER'S SECTION
MAKING AN ELECTROMAGNET

GOALS AND OBJECTIVES:

1. To learn that magnetism can be produced from electricity.

2. To learn that a flow of electrons along a wire produces a magnetic field around the wire.

3. To learn that in order to produce magnetic effects without a magnet, one needs electricity.

4. To learn that electricity and magnetism are closely related to each other.

5. To learn that an electromagnet retains its magnetic properties only when the electricity is turned on.

6. To learn that electromagnets are temporary magnets.

DISCUSSION BEFORE THE LABORATORY ACTIVITY:

It is very important to stress the need for wearing an apron and safety goggles.

It was discovered a long time ago that lodestone, a rock found naturally on earth, attracts small pieces of iron. Lodestone is now known to chemists as a compound of iron oxide. Found near the ancient city of Magnesia, Chinese sailors discovered that rubbing a lodestone with an iron needle made the needle attract iron pieces. They also discovered that such a needle, when suspended freely, always points north.

A compass contains a magnetized needle that can turn freely. The needle always points north, no matter how the compass is turned.

A magnet is a substance that attracts iron. Iron, steel, cobalt, and nickel are all magnetic. Metals such as copper, zinc, and gold are not magnetic. Nonmetals such as paper and glass are not magnetic.

The earth itself acts as a magnet. The geographic north and south poles of the earth are on the axis of rotation. The magnetic north pole and south pole do not coincide with the geographic north pole and south pole. The magnetization axis makes an angle of about 15 degrees with the earth's axis. The entire earth is assumed to be a magnet with two poles. It is assumed that the north pole of the earth's magnet is distributed over the entire southern hemisphere, and the south pole of the earth's magnet is distributed over the entire northern hemisphere. Thus the north pole of the earth's magnet is directed toward the general geographic south pole and the south pole of the earth's magnet is directed towards the general geographic north pole.

It is possible to use earth's magnetism to produce other magnets.

Temporary magnets are made using soft iron. Permanent magnets of differing size, shape and strength are made of steel. The most common ones are bar magnets, horseshoe magnets, and circular magnets. Very powerful ceramic magnets, alnico (an alloy of iron, aluminum, nickel and cobalt) magnets, and Neo (an alloy of neodymium, iron, and boron) magnets are also available commercially. Ceramic magnets are stronger than alnico magnets. Neo magnets are much stronger than ceramic magnets.

An electromagnet is a temporary magnet with a core made of soft iron. It will behave like a magnet only when electricity is passing through the coil surrounding the iron core. Very powerful electromagnets are made using electricity. The great advantage of the electromagnet is that it will act as a magnet only when the electricity is turned on. It loses its magnetic properties when the electricity is turned off.

When the wire is coiled, the flow of electrons through the coil makes strong magnetic fields. By induction, the iron core inside the coil will act like a magnet with north and south poles when electricity is flowing through the coil. The poles of the electromagnet can be switched by changing the direction of electricity flowing through the coil. The invention of the electric motor was based on this ability of an electromagnet to switch the poles.

Moving electric charges produce magnetic forces or fields around them. Electricity and magnetism are closely related to each other. Like electrical charges, such as two positive charges or two negative charges, repel each other. Like poles of two magnets, such as two north poles or two south poles, also repel each other. Unlike electrical charges, such as a positive charge and a negative charge, attract each other. Unlike poles of two magnets, such as a north pole and a south pole, also attract each other. Electricity can be produced from magnetism and magnetism can be produced from electricity.

The connection between electricity and magnetism was first observed in 1819 by the Danish scientist, Hans Christian Oersted. He observed that a current carrying wire deflected the needle of a compass. The forces exerted by the magnetic field produced from electricity and from permanent magnets are the same in all aspects.

In this activity, two 1.5 volts dry "D" cells are used. You can get "D" cell connectors from an electronic shop. This connector provides color–coded connections, eliminating the need for taping the cells in a series. Connecting the negative terminal of a cell to the positive terminal of a second cell will give a series connection.

TEACHER'S SECTION
NUMBER OF TURNS IN THE COIL OF AN ELECTROMAGNET

GOALS AND OBJECTIVES:

1. To learn that magnetism can be produced from electricity.
2. To learn that a flow of electrons along a wire produces a magnetic field around the wire.
3. To learn that in order to produce magnetic effects without a magnet, one needs electricity.
4. To learn that electricity and magnetism are closely related to each other.
5. To learn that an electromagnet retains its magnetic properties only when the electricity is turned on.
6. To learn that electromagnets are temporary magnets.
7. To learn that the strength of an electromagnet depends on the number of turns in the coil.
8. To observe that the strength of the electromagnet increases with increasing number of turns in the coil.

DISCUSSION BEFORE THE LABORATORY ACTIVITY:

It is very important to stress the need for wearing an apron and safety goggles.

It was discovered a long time ago that lodestone, a rock found naturally on earth, attracts small pieces of iron. Lodestone is now known to chemists as a compound of iron oxide. Found near the ancient city of Magnesia, Chinese sailors discovered that rubbing a lodestone with an iron needle made the needle attract iron pieces. They also discovered that such a needle, when suspended freely, always points north.

A compass contains a magnetized needle that can turn freely. The needle always points north, no matter how the compass is turned.

A magnet is a substance that attracts iron. Iron, steel, cobalt, and nickel are all magnetic. Metals such as copper, zinc, and gold are not magnetic. Nonmetals such as paper and glass are not magnetic.

The earth itself acts as a magnet. The geographic north and south poles of the earth are on the axis of rotation. The magnetic north pole and south pole do not coincide with the geographic north pole and south pole. The magnetization axis makes an angle of about 15 degrees with the earth's axis. The entire earth is assumed to be a magnet with two poles. It is assumed that the north pole of the earth's magnet is distributed over the entire southern hemisphere, and the south pole of the earth's magnet is distributed over the entire northern hemisphere. Thus the north pole of the earth's magnet is directed toward the general geographic south pole and the south pole of the earth's magnet is directed towards the general geographic north pole.

It is possible to use earth's magnetism to produce other magnets.

Temporary magnets are made using soft iron. Permanent magnets of differing size, shape and strength are made of steel. The most common ones are bar magnets, horseshoe magnets, and circular magnets. Very powerful ceramic magnets, alnico (an alloy of iron, aluminum, nickel and cobalt) magnets, and Neo (an alloy of neodymium, iron, and boron) magnets are also available commercially. Ceramic magnets are stronger than alnico magnets. Neo magnets are much stronger than ceramic magnets.

An electromagnet is a temporary magnet with a core made of soft iron. It will behave like a magnet only when electricity is passing through the coil surrounding the iron core. Very powerful electromagnets are made using electricity. The great advantage of the electromagnet is that it will act as a magnet only when the electricity is turned on. It loses its magnetic properties when the electricity is turned off.

The strength of an electromagnet increases with an increasing number of turns in the coil, with increasing current passing through the coil, and by using a permeable material as a core inside the coil.

Moving electric charges produce magnetic forces or fields around them. Electricity and magnetism are closely related to each other. Like electrical charges, such as two positive charges or two negative charges, repel each other. Like poles of two magnets, such as two north poles or two south poles, also repel each other. Unlike electrical charges, such as a positive charge and a negative charge, attract each other. Unlike poles of two magnets, such as a north pole and a south pole, also attract each other. Electricity can be produced from magnetism and magnetism can be produced from electricity.

The connection between electricity and magnetism was first observed in 1819 by the Danish scientist, Hans Christian Oersted. He observed that a current carrying wire deflected the needle of a compass. The forces exerted by the magnetic field produced from electricity and from permanent magnets are the same in all aspects.

In this activity, two 1.5 volts dry "D" cells are used. You can get "D" cell connectors from an electronic shop. This connector provides color–coded connections, eliminating the need for taping the cells in a series. Connecting the negative terminal of a cell to the positive terminal of a second cell will give a series connection.

TEACHER'S SECTION
THE AMOUNT OF CURRENT IN THE COIL OF AN ELECTROMAGNET

GOALS AND OBJECTIVES:

1. To learn that magnetism can be produced from electricity.

2. To learn that a flow of electrons along a wire produces a magnetic field around the wire.

3. To learn that in order to produce magnetic effects without a magnet, one needs electricity.

4. To learn that electricity and magnetism are closely related to each other.

5. To learn that an electromagnet retains its magnetic properties only when the electricity is turned on.

6. To learn that electromagnets are temporary magnets.

7. To learn that the strength of an electromagnet depends on the amount of current passing through the coil.

8. To observe that the strength of the electromagnet increases with increasing current passing through the coil.

DISCUSSION BEFORE THE LABORATORY ACTIVITY:

It is very important to stress the need for wearing an apron and safety goggles.

It was discovered a long time ago that lodestone, a rock found naturally on earth, attracts small pieces of iron. Lodestone is now known to chemists as a compound of iron oxide. Found near the ancient city of Magnesia, Chinese sailors discovered that rubbing a lodestone with an iron needle made the needle attract iron pieces. They also discovered that such a needle, when suspended freely, always points north.

A compass contains a magnetized needle that can turn freely. The needle always points north, no matter how the compass is turned.

A magnet is a substance that attracts iron. Iron, steel, cobalt, and nickel are all magnetic. Metals such as copper, zinc, and gold are not magnetic. Nonmetals such as paper and glass are not magnetic.

The earth itself acts as a magnet. The geographic north and south poles of the earth are on the axis of rotation. The magnetic north pole and south pole do not coincide with the geographic north pole and south pole. The magnetization axis makes an angle of about 15 degrees with the earth's axis. The entire earth is assumed to be a magnet with two poles. It is assumed that the north pole of the earth's magnet is distributed over the entire southern hemisphere, and the south pole of the earth's magnet is distributed over the entire northern hemisphere. Thus the north pole of the earth's magnet is directed toward the general geographic south pole and the south pole of the earth's magnet is directed towards the general geographic north pole.

It is possible to use earth's magnetism to produce other magnets.

Temporary magnets are made using soft iron. Permanent magnets of differing size, shape and strength are made of steel. The most common ones are bar magnets, horseshoe magnets, and circular magnets. Very powerful ceramic magnets, alnico (an alloy of iron, aluminum, nickel and cobalt) magnets, and Neo (an alloy of neodymium, iron, and boron) magnets are also available commercially. Ceramic magnets are stronger than alnico magnets. Neo magnets are much stronger than ceramic magnets.

An electromagnet is a temporary magnet with a core made of soft iron. It will behave like a magnet only when electricity is passing through the coil surrounding the iron core. Very powerful electromagnets are made using electricity. The great advantage of the electromagnet is that it will act as a magnet only when the electricity is turned on. It loses its magnetic properties when the electricity is turned off.

When the wire is coiled, the flow of electrons through the coil makes strong magnetic fields. By induction, the iron core inside the coil will act like a magnet with north and south poles when electricity is flowing through the coil. The poles of the electromagnet can be switched by changing the direction of electricity flowing through the coil. The invention of the electric motor was based on this ability of an electromagnet to switch the poles.

The strength of an electromagnet increases with increasing number of turns in the coil, with increasing current passing through the coil, and by using a permeable material as a core inside the coil.

Moving electric charges produce magnetic forces or fields around them. Electricity and magnetism are closely related to each other. Like electrical charges, such as two positive charges or two negative charges, repel each other. Like poles of two magnets, such as two north poles or two south poles, also repel each other. Unlike electrical charges, such as a positive charge and a negative charge, attract each other. Unlike poles of two magnets, such as a north pole and a south pole, also attract each other. Electricity can be produced from magnetism and magnetism can be produced from electricity.

The connection between electricity and magnetism was first observed in 1819 by the Danish scientist, Hans Christian Oersted. He observed that a current carrying wire deflected the needle of a compass. The forces exerted by the magnetic field produced from electricity and from permanent magnets are the same in all aspects.

In this activity, two 1.5 volts dry "D" cells are used. You can get "D" cell connectors from an electronic shop. This connector provides color–coded connections, eliminating the need for taping the cells in a series. Connecting the negative terminal of the first cell to the positive terminal of the second cell, connecting the negative terminal of the second cell to the positive terminal of the third cell, and so on will give a series connection.

TEACHER'S SECTION
USE OF PERMEABLE MATERIALS TO STRENGTHEN AN ELECTROMAGNET

GOALS AND OBJECTIVES:

1. To learn that magnetism can be produced from electricity.
2. To learn that a flow of electrons along a wire produces a magnetic field around the wire.
3. To learn that in order to produce magnetic effects without a magnet, one needs electricity.
4. To learn that electricity and magnetism are closely related to each other.
5. To learn that an electromagnet retains its magnetic properties only when the electricity is turned on.
6. To learn that electromagnets are temporary magnets.
7. To observe that the strength of an electromagnet increases by using a permeable material as a core inside the coil.

DISCUSSION BEFORE THE LABORATORY ACTIVITY:

It is very important to stress the need for wearing an apron and safety goggles.

It was discovered a long time ago that lodestone, a rock found naturally on earth, attracts small pieces of iron. Lodestone is now known to chemists as a compound of iron oxide. Found near the ancient city of Magnesia, Chinese sailors discovered that rubbing a lodestone with an iron needle made the needle attract iron pieces. They also discovered that such a needle, when suspended freely, always points north.

A compass contains a magnetized needle that can turn freely. The needle always points north, no matter how the compass is turned.

A magnet is a substance that attracts iron. Iron, steel, cobalt, and nickel are all magnetic. Metals such as copper, zinc, and gold are not magnetic. Nonmetals such as paper and glass are not magnetic.

The earth itself acts as a magnet. The geographic north and south poles of the earth are on the axis of rotation. The magnetic north pole and south pole do not coincide with the geographic north pole and south pole. The magnetization axis makes an angle of about 15 degrees with the earth's axis. The entire earth is assumed to be a magnet with two poles. It is assumed that the north pole of the earth's magnet is distributed over the entire southern hemisphere, and the south pole of the earth's magnet is distributed over the entire northern hemisphere. Thus the north pole of the earth's magnet is directed toward the general geographic south pole and the south pole of the earth's magnet is directed towards the general geographic north pole.

It is possible to use earth's magnetism to produce other magnets.

Temporary magnets are made using soft iron. Permanent magnets of differing size, shape and strength are made of steel. The most common ones are bar magnets, horseshoe magnets, and circular magnets. Very powerful ceramic magnets, alnico (an alloy of iron, aluminum, nickel and cobalt) magnets, and Neo (an alloy of neodymium, iron, and boron) magnets are also available commercially. Ceramic magnets are stronger than alnico magnets. Neo magnets are much stronger than ceramic magnets.

An electromagnet is a temporary magnet with acore made of soft iron. It will behave like a magnet only when electricity is passing through the coil surrounding the iron core. Very powerful electromagnets are made using electricity. The great advantage of the electromagnet is that it will act as a magnet only when the electricity is turned on. It loses its magnetic properties when the electricity is turned off.

When the wire is coiled, the flow of electrons through the coil makes strong magnetic fields. By induction, the iron core inside the coil will act like a magnet with north and south poles when electricity is flowing through the coil. The poles of the electromagnet can be switched by changing the direction of electricity flowing through the coil. The invention of the electric motor was based on this ability of an electromagnet to switch the poles.

The strength of an electromagnet increases with increasing number of turns in the coil, with increasing current passing through the coil, and by using a permeable material as a core inside the coil.

Moving electric charges produce magnetic forces or fields around them. Electricity and magnetism are closely related to each other. Like electrical charges, such as two positive charges or two negative charges, repel each other. Like poles of two magnets, such as two north poles or two south poles, also repel each other. Unlike electrical charges, such as a positive charge and a negative charge, attract each other. Unlike poles of two magnets, such as a north pole and a south pole, also attract each other.

Electricity can be produced from magnetism and magnetism can be produced from electricity.

The connection between electricity and magnetism was first observed in 1819 by the Danish scientist, Hans Christian Oersted. He observed that a current carrying wire deflected the needle of a compass. The forces exerted by the magnetic field produced from electricity and from permanent magnets are the same in all aspects.

In this activity, two 1.5 volts dry "D" cells are used. You can get "D" cell connectors from an electronic shop. This connector provides color–coded connections, eliminating the need for taping the cells in a series. Connecting the negative terminal of the first cell to the positive terminal of the second cell will give a series connection.

HAINES MEDIA CENTER